Dictionary of Mathematics Terms

Second Edition

K

Dictionary of Mathematics Terms

Second Edition

Douglas Downing, Ph.D.

School of Business and Economics
Seattle Pacific University

BARRON'S

Clark, and Robert Downing for their special help.

All inquiries should be addressed to:
Barron's Educational Series, Inc.
250 Wireless Boulevard
Hauppauge, New York 11788

Library of Congress Catalog Card No. 95-12039

International Standard Book No. 0-8120-3097-4

Library of Congress Cataloging in Publication Data

Downing, Douglas
   Dictionary of mathematics terms/Douglas Downing.–
      2nd ed.
         p.   cm.   –(Barron's professional guides)
      ISBN 0-8120-3097-4
   1. Mathematics — dictionaries.   I. Title.
QA5.D69   1995
510'.3–dc20   95-12039
                                                    CIP

PRINTED IN THE UNITED STATES OF AMERICA

 678   6550   98765432

CONTENTS

PREFACE

Mathematics consists of rigorous abstract reasoning. At first, it can be intimidating; but learning about math can help you appreciate its great practical usefulness and even its beauty—both for the visual appeal of geometric forms and the concise elegance of symbolic formulas expressing complicated ideas.

Imagine that you are to build a bridge, or a radio, or a bookcase. In each case you should plan first, before beginning to build. In the process of planning you will develop an abstract model of the finished object—and when you do that, you are doing mathematics.

The purpose of this book is to collect in one place reference information that is valuable for students of mathematics and for persons with careers that use math. The book covers mathematics that is studied in high school and the early years of college. These are some of the general subjects that are included:

Arithmetic: the properties of numbers and the four basic operations: addition, subtraction, multiplication, division.

Algebra: the first step to abstract symbolic reasoning. In algebra we study operations on symbols (usually letters) that stand for numbers. This makes it possible to develop many general results. It also saves work because it is possible to derive symbolic formulas that will work for whatever numbers you put in; this saves you from having to derive the solution again each time you change the numbers.

Geometry: the study of shapes. Geometry has great visual appeal, and it is also important because it is an example of a rigorous logical system where theorems are proved on the basis of postulates and previously proved theorems.

Analytic Geometry: where algebra and geometry come together as algebraic formulas are used to describe geometric shapes.

Trigonometry: the study of triangles, but also much more. Trigonometry focuses on six functions defined in terms of the sides of right angles (sine, cosine, tangent, secant, cosecant, cotangent) but then it takes many surprising turns. For example, oscillating phenomena such as pendulums, springs, water waves, light waves, sound waves, and electronic circuits can all be described in terms of trigonometric functions. If you program a computer to picture an object on the screen, and you wish to rotate it to view it from a different angle, you will use trigonometry to calculate the rotated position.

Calculus: the study of rates of change, and much more. Begin by asking these questions: how much does one value change when another value changes? How fast does an object move? How steep is a slope? These problems can be solved by calculating the derivative, which also allows you to answer the question: what is the highest or lowest value? Reverse this process to calculate an integral, and something amazing happens: integrals can also be used to calculate areas, volumes, arc lengths, and other quantities. A first course in calculus typically covers the calculus of one variable; this book also includes some topics in multi-variable calculus, such as partial derivatives and double integrals.

Probability and Statistics: the study of chance phenomena, and how that study can be applied to the analysis of data.

Logic: the study of reasoning.

People: several mathematicians who have made major contributions throughout history are included.

A list of entries by subject category is included at the front of the book. This list will help you find more information on a particular branch of mathematics. The appendix includes several useful tables.

Demonstrations of important theorems, such as the Pythagorean theorem and the quadratic formula, are included. Many entries contain cross references indicating where to find background information or further applications of the topic. A list of symbols at the beginning of the book helps the reader identify unfamiliar symbols.

Douglas Downing, Ph.D.
Seattle, Washington
1995

LIST OF SYMBOLS

Algebra

$=$	equals
$\neq$	is not equal
$\approx$	is approximately equal
$>$	is greater than
$\geq$	is greater than or equal to
$<$	is less than
$\leq$	is less than or equal to
$+$	addition
$-$	subtraction
$\times, \cdot$	multiplication
$\div, /$	division
$\sqrt{}$	square root; radical symbol
$\sqrt[n]{}$	nth root
$!$	factorial
$_nC_j, \binom{n}{j}$	number of combinations of n things taken j at a time; also the binomial theorem coefficient
$_nP_j$	number of permutations of n things taken j at a time
$\lvert x \rvert$	absolute value of x
∞	infinity
$\begin{vmatrix} a & b \\ c & d \end{vmatrix}$	determinant of a matrix

Greek Letters

π	pi $(= 3.14159\ldots)$
Δ	delta (upper case), represents change in
δ	delta (lower case)
Σ	sigma (upper case), represents summation
σ	sigma (lower case), represents standard deviation

θ	theta (used for angles)
ϕ	phi (used for angles)
μ	mu, represents mean
ϵ	epsilon
χ	chi
ρ	rho (correlation coefficient)
λ	lambda

Calculus

Δx	increment of x
$y', \dfrac{dy}{dx}$	derivative of y with respect to x
$y'', \dfrac{d^2y}{dx^2}$	second derivative of y with respect to x
$\dfrac{\partial y}{\partial x}$	partial derivative of y with respect to x
$\rightarrow$	approaches
lim	limit
e	base of natural logarithms; $e = 2.71828\ldots$
$\int$	integral symbol
$\int f(x)dx$	indefinite integral
$\int_a^b f(x)dx$	definite integral

Geometry

$\circ$	degrees
$\llcorner$	perpendicular
$\perp$	perpendicular, as in $\overline{AB} \perp \overline{DC}$
$\angle$	angle
$\triangle$	triangle, as in $\triangle ABC$
$\cong$	congruent
$\sim$	similar

$\parallel$	parallel, as in $\overline{AB} \parallel \overline{CD}$
$\frown$	arc, as in $\overset{\frown}{AB}$
$-$	line segment, as in $\overline{AB}$
$\leftrightarrow$	line, as in $\overset{\leftrightarrow}{AB}$
$\rightarrow$	ray, as in $\overset{\rightarrow}{AB}$

Vectors

$\parallel \mathbf{a} \parallel$	length of vector $\mathbf{a}$
$\mathbf{a} \cdot \mathbf{b}$	dot product
$\mathbf{a} \times \mathbf{b}$	cross product
∇f	gradient
$\nabla \cdot \mathbf{f}$	divergence
$\nabla \times \mathbf{f}$	curl

Set Notation

$\{\ \}$	braces (indicating membership in a set)
$\cap$	intersection
$\cup$	union
$\emptyset$	empty set

Logic

$\rightarrow$	implication, as in $a \rightarrow b$ (IF a THEN b)
$\sim p$	the negation of a proposition p
$\wedge$	conjunction (AND)
$\vee$	disjunction (OR)
IFF, $\leftrightarrow$	equivalence, (IF AND ONLY IF)
$\forall x$	universal quantifier (means "For all x...")
$\exists x$	existential quantifier (means "There exists an x...")

GUIDE TO SELECTED TERMS BY CATEGORY

Algebra
absolute value
algebra
binomial
binomial theorem
common logarithm
completing the square
complex number
conditional equation
conjugate
cube root
difference of two squares
equation
factor
factor theorem
factoring
function
fundamental theorem
    of algebra
geometric series
imaginary number
independent variable
inverse function
like terms
linear equation
logarithm
monomial
multinomial
polynomial
quadratic equation
quadratic equation,
    2 unknowns
quadratic formula
radical
rational root theorem
rationalizing the
    denominator
simultaneous equations
solution
square root
summation notation
synthetic division
system of equations
system of inequalities
term
trinomial
variable

Analytic Geometry
abscissa
analytic geometry
Cartesian coordinates
circle
conic sections
coordinates
directrix
eccentricity
ellipse
hyperbola
intercept

ordinate
origin
parabola
polar coordinates
rectangular coordinates
translation
x-axis
y-axis

Arithmetic
addition
additive identity
additive inverse
arithmetic mean
associative property
average
closure property
commutative property
complex fraction
composite number
counting numbers
decimal numbers
denominator
difference
digit
distributive property
dividend
division
divisor
Eratosthenes sieve
Euclid's algorithm
even number
exponent
fraction

fundamental theorem
    of arithmetic
greatest common factor
improper fraction
inequality
infinity
integers
irrational number
least common
    denominator
least common multiple
multiplicand
multiplication
multiplicative identity
multiplicative inverse
natural numbers
negation
negative
number
numerator
odd number
positive number
power
prime factors
prime number
product
proper fraction
quotient
rational number
real numbers
reciprocal
remainder
repeating decimal
subtraction

chord
circle
circumscribed
collinear
compass
complementary angles
cone
congruent
coplanar
corresponding angles
corresponding sides
cube
cylinder
decagon
diagonal
diameter
dihedral angle
distance
dodecahedron
edge
equilateral triangle
Euclidian geometry
face
frustum
geometric construction
geometry
heptagon
hexagon
hexahedron
icosahedron
inscribed
isosceles triangle
line
line segment

midpoint
non-Euclidian geometry
oblique angle
oblique triangle
obtuse angle
obtuse triangle
octagon
octahedron
parallel
parallelepiped
parallelogram
pentagon
perimeter
perpendicular
plane
polygon
polyhedron
prism
projection
protractor
pyramid
Pythagorean theorem
quadrilateral
radius
ray
rectangle
regular polygon
regular polyhedron
rhombus
right angle
right circular cone
right circular cylinder
right triangle
scalene triangle

sector
segment
similar
skew
slope
solid
sphere
square
supplementary
surface
surface area
symmetric
tetrahedron
trapezoid
triangle
truncated cone
truncated pyramid
vertex
vertical angles
volume

Logic
and
antecedent
argument
axiom
biconditional statement
Boolean algebra
compound sentence
conclusion
conditional statement
conjecture
conjunction
consequent

contradiction
contrapositive
converse
corollary
deduction
de Morgan's laws
disjunction
equivalent
existential quantifier
false
Gödel's incompleteness
    theorem
if
implication
indirect proof
induction
lemma
logic
mathematical induction
necessary
not
or
postulate
premise
proof
sentence
sufficient
syllogism
tautology
then
theorem
true
truth table
undefined term

Trigonometry

.

A

ABELIAN GROUP See **group.**

ABSCISSA Abscissa means x-coordinate. The abscissa of the point (a, b) in Cartesian coordinates is a. For contrast, see **ordinate.**

ABSOLUTE VALUE The absolute value of a real number a, written as $|a|$, is:

$$|a| = a \text{ if } a \geq 0$$
$$|a| = -a \text{ if } a < 0$$

Absolute values are always positive or zero. If all the real numbers are represented on a number line, you can think of the absolute value of a number as being the distance from zero to that number. You can find absolute values by leaving positive numbers alone and ignoring the sign of negative numbers. For example, $|17| = 17, |-105| = 105, |0| = 0$

The absolute value of a complex number $a + bi$ is $\sqrt{a^2 + b^2}$.

ACCELERATION The acceleration of an object measures the rate of change in its velocity. For example, if a car increases its velocity from 0 to 24.6 meters per second (55 miles per hour) in 12 seconds, its acceleration was 2.05 meters per second per second, or 2.05 meters/second-squared.

If $x(t)$ represents the position of an object moving in one dimension as a function of time, then the first derivative, dx/dt, represents the velocity of the object, and the second derivative, d^2x/dt^2, represents the acceleration. Newton found that, if F represents the force acting on an object and m represents its mass, the acceleration (a) is determined from the formula $F = ma$.

ACUTE ANGLE An acute angle is a positive angle smaller than a 90° angle.

ACUTE TRIANGLE An acute triangle is a triangle wherein each of the three angles is smaller than a 90° angle. For contrast, see **obtuse triangle**.

ADDITION Addition is the operation of combining two numbers to form a sum. For example, $3 + 4 = 7$. Addition satisfies two important properties: the commutative property, which says that

$$a + b \;=\; b + a \text{ for all } a \text{ and } b$$

and the associative property, which says that

$$(a + b) + c \;=\; a + (b + c) \text{ for all } a, \ b, \text{ and } c.$$

ADDITIVE IDENTITY The number zero is the additive identity element, because it satisfies the property that the addition of zero does not change a number: $a + 0 = a$ for all a.

ADDITIVE INVERSE The sum of a number and its additive inverse is zero. The additive inverse of a (written as $-a$) is also called the negative or the opposite of a: $a + (-a) = 0$. For example, -1 is the additive inverse of 1, and 10 is the additive inverse of -10.

ADJACENT ANGLES Two angles are adjacent if they share the same vertex and have one side in common between them.

ALGEBRA Algebra is the study of properties of operations carried out on sets of numbers. Algebra is a generalization of arithmetic in which symbols, usually letters, are used to stand for numbers. The structure of algebra is based upon axioms (or postulates), which are statements that are assumed to be true. Some algebraic axioms include the transitive axiom:

$$\text{if } a = b \text{ and } b = c, \text{ then } a = c$$

and the associative axiom of addition:

$$(a + b) + c = a + (b + c)$$

These axioms are then used to prove theorems about the properties of operations on numbers.

A common problem in algebra involves solving conditional equations—in other words, finding the values of an unknown that make the equation true. An equation of the general form $ax + b = 0$, where x is unknown and a and b are known, is called a **linear equation.** An equation of the general form $ax^2 + bx + c = 0$ is called a **quadratic equation.** For equations involving higher powers of x, see **polynomial.** For situations involving more than one equation with more than one unknown, see **simultaneous equations.**

This article has described elementary algebra. Higher algebra involves the extension of symbolic reasoning into other areas that are beyond the scope of this book.

ALGORITHM An algorithm is a sequence of instructions that tell how to accomplish a task. An algorithm must be specified exactly, so that there can be no doubt about what to do next, and it must have a finite number of steps.

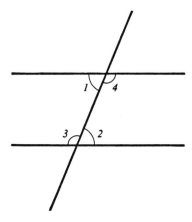

Figure 1 Alternate interior angles

AL-KHWARIZMI Muhammad Ibn Musa Al-Khwarizmi (c 780 AD to c 850 AD) was a Muslim
mathematician whose works introduced our modern
numerals (the Hindu-arabic numerals) to Europe,
and the title of his book *Kitab al-jabr wa al-mu-
qabalah* provided the source for the term algebra. His
name is the source for the term algorithm.

ALTERNATE INTERIOR ANGLES When a
transversal cuts two lines, it forms two pairs of alternate interior angles. In figure 1, ∠1 and ∠2 are a
pair of alternate interior angles, and ∠3 and ∠4 are
another pair. A theorem in Euclidian geometry says
that, when a transversal cuts two parallel lines, any
two alternate interior angles will equal each other.

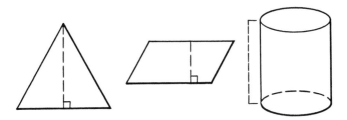

Figure 2 Altitudes

ALTERNATING SERIES An alternating series is a series in which every term has the opposite sign from the preceding term. For example, $x - x^3/3! + x^5/5! - x^7/7! + x^9/9! - ...$ is an alternating series.

ALTERNATIVE HYPOTHESIS The alternative hypothesis is the hypothesis that states, "The null hypothesis is false." (See **hypothesis testing**.)

ALTITUDE The altitude of a plane figure is the distance from one side, called the base, to the farthest point. The altitude of a solid is the distance from the plane containing the base to the highest point in the solid. In figure 2, the dotted lines show the altitude of a triangle, of a parallelogram, and of a cylinder.

AMBIGUOUS CASE The term "ambiguous case" refers to a situation in which you know the lengths of two sides of a triangle and you know one of the angles (other than the angle between the two sides of known lengths). If the known angle is less than $90°$, it may not be possible to solve for the length of the third side or for the sizes of the other two angles. In figure 3, side AB of the upper triangle is the same

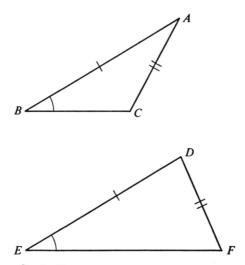

Figure 3 Ambiguous case

length as side DE of the lower triangle, side AC is
the same length as side DF, and angle B is the same
size as angle E. However, the two triangles are quite
different. (See also **solving triangles**.)

AMPLITUDE The amplitude of a periodic function
is one-half the difference between the largest possible
value of the function and the smallest possible value.
For example, for $y = \sin x$, the largest possible value
of y is 1 and the smallest possible value is -1, so
the amplitude is 1. In general, the amplitude of the
function $y = A \sin x$ is $|A|$.

ANALOG An analog system is a system in which
numbers are represented by a device that can vary

continuously. For example, a slide rule is an example of an analog calculating device, because numbers are represented by the distance along a scale. If you could measure the distances perfectly accurately, then a slide rule would be perfectly accurate; however, in practice the difficulty of making exact measurements severely limits the accuracy of analog devices. Other examples of analog devices include clocks with hands that move around a circle, thermometers in which the temperature is indicated by the height of the mercury, and traditional records in which the amplitude of the sound is represented by the height of a groove. For contrast, see **digital.**

ANALYSIS Analysis is the branch of mathematics that studies limits and convergence; calculus is a part of analysis.

ANALYSIS OF VARIANCE Analysis of variance (ANOVA) is a procedure used to test the hypothesis that three or more different samples were all selected from populations with the same mean. The method is based on a test statistic:

$$F = \frac{nS*^2}{S^2}$$

where n is the number of members in each sample, $S*^2$ is the variance of the sample averages for all of the groups, and S^2 is the average variance for the groups. If the null hypothesis is true and the population means actually are all the same, this statistic will have an F distribution with $(m-1)$ and $m(n-1)$ degrees of freedom, where m is the number of samples. If the value of the test statistic is too large, the null hypothesis is rejected. (See **hypothesis test-**

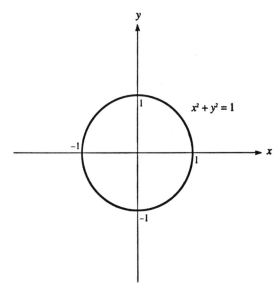

Figure 4 Equation of circle

ing.) Intuitively, a large value of $S*^2$ means that the observed sample averages are spread further apart, thereby making the test statistic larger and the null hypothesis less likely to be accepted.

ANALYTIC GEOMETRY Analytic geometry is the branch of mathematics that uses algebra to help in the study of geometry. It helps you understand algebra by allowing you to draw pictures of algebraic equations, and it helps you understand geometry by allowing you to describe geometric figures by means of algebraic equations. Analytic geometry is based

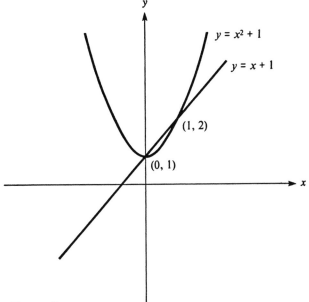

Figure 5

on the fact that there is a one-to-one correspondence between the set of real numbers and the set of points on a number line. Any point in a plane can be described by an ordered pair of numbers (x, y). (See **Cartesian coordinates**.) The graph of an equation in two variables is the set of all points in the plane that are represented by an ordered pair of numbers that make the equation true. For example, the graph of the equation $x^2 + y^2 = 1$ is a circle with its center at the origin and a radius of 1. (See figure 4.)

A linear equation is an equation in which both x and y occur to the first power, and there are no terms containing xy. Its graph will be a straight line. (See **linear equation**.) When either x or y (or both) is raised to the second power, some interesting curves can result. (See **conic sections; quadratic equations, two unknowns**.) When higher powers of the variable are used, it is possible to draw curves with many changes of direction. (See **polynomial**.)

Graphs can also be used to illustrate the solutions for systems of equations. If you are given two equations in two unknowns, draw the graph of each equation. The places where the two curves intersect will be the solutions to the system of equations. (See **simultaneous equations**.) Figure 5 shows the solution to the system of equations $y = x+1, y = x^2+1$.

Although Cartesian, or rectangular, coordinates are the most commonly used, it is sometimes helpful to use another type of coordinates known as **polar coordinates**.

AND The word "AND" is a connective word used in logic. The sentence "p AND q" is true only if both sentence p as well as sentence q are true. The operation of AND is illustrated by the truth table:

p	q	p AND q
T	T	T
T	F	F
F	T	F
F	F	F

AND is often represented by the symbol $\wedge$ or &. An AND sentence is also called a *conjunction*. (See **logic; Boolean algebra**.)

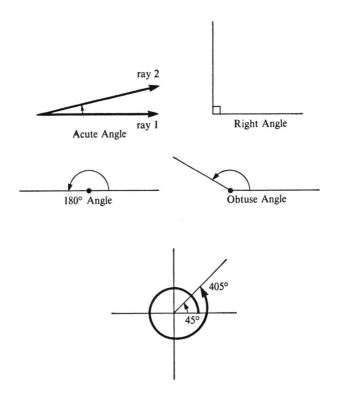

Figure 6 Angles

ANGLE An angle is the union of two rays with a common endpoint. If the two rays point in the same direction, then the angle between them is zero. Suppose that ray 1 is kept fixed, and ray 2 is pivoted counterclockwise about its endpoint. The measure of an angle is a measure of how much ray 2 has been rotated. If ray 2 is rotated a complete turn, so that it again points in the same direction as ray 1, we say that it has been turned 360 degrees (written as 360°) or 2π radians. A half turn measures 180°, or π radians. A quarter turn, forming a square corner, measures 90°, or $\pi/2$ radians. Such an angle is also known as a right angle.

An angle smaller than a 90° angle is called an *acute angle*. An angle larger than a 90° angle but smaller than a 180° angle is called an *obtuse angle*. See figure 6.

For some mathematical purposes it is useful to allow for general angles that can be larger than 360°, or even negative. A general angle still measures the amount that ray 2 has been rotated in a counterclockwise direction. A 720° angle (meaning two full rotations) is the same as a 360° angle (one full rotation), which in turn is the same as a 0° angle (no rotation). Likewise, a 405° angle is the same as a 45° angle (since $405 - 360 = 45$). (See figure 6.)

A negative angle is the amount that ray 2 has been rotated in a clockwise direction. A −90° angle is the same as a 270° angle.

Conversions between radian and degree measure can be made by multiplication:

$$(degree\ measure) = \frac{180}{\pi} \times (radian\ measure)$$

$$(radian\ measure) = \frac{\pi}{180} \times (degree\ measure)$$

One radian is about 57°.

ANGLE OF DEPRESSION The angle of depression for an object below your line of sight is the angle whose vertex is at your position, with one side being a horizontal ray in the same direction as the object

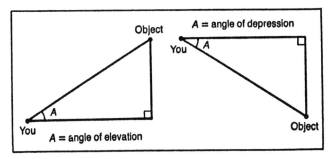

Figure 7

and the other side being the ray from your eye passing through the object. (See figure 7.)

ANGLE OF ELEVATION The angle of elevation for an object above your line of sight is the angle whose vertex is at your position, with one side being a horizontal ray in the same direction as the object and the other side being the ray from your eye passing through the object. (See figure 7.)

ANGLE OF INCIDENCE When a light ray strikes a surface, the angle between the ray and the normal to the surface is called the angle of incidence. (The normal is the line perpendicular to the surface.) If it is a reflective surface, such as a mirror, then the angle formed by the light ray as it leaves the surface is called the angle of reflection. A law of optics states that the angle of reflection is equal to the angle of incidence. (See figure 8.)

See **Snell's law** for a discussion of what happens when the light ray travels from one medium to another, such as from air to water or glass.

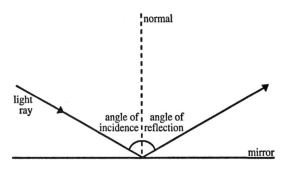

Figure 8

ANGLE OF REFLECTION See **angle of incidence**.

ANGLE OF REFRACTION See **Snell's law**.

ANTECEDENT The antecedent is the "if" part of an "if/then" statement. For example, in the statement "If he likes pizza, then he likes cheese," the antecedent is the clause "he likes pizza."

ANTIDERIVATIVE An antiderivative of a function $f(x)$ is a function $F(x)$ whose derivative is $f(x)$ (that is, $dF(x)/dx = f(x)$). $F(x)$ is also called the **indefinite integral** of $f(x)$.

ANTILOGARITHM If $y = \log_a x$, (in other words, $x = a^y$), then x is the antilogarithm of y to the base a. (See **logarithm**.)

APOLLONIUS Apollonius of Perga (262 BC to 190 BC), a mathematician who studied in Alexandria under pupils of Euclid, wrote works that extended Euclid's work in geometry, particularly focusing on conic sections.

APOTHEM The apothem of a regular polygon is the distance from the center of the polygon to one of the sides of the polygon, in the direction that is perpendicular to that side.

ARC An arc of a circle is the set of points on the circle that lie in the interior of a particular central angle. Therefore an arc is a part of a circle. The degree measure of an arc is the same as the degree measure of the angle that defines it. If θ is the degree measure of an arc and r is the radius, then the length of the arc is $2\pi r\theta/360$. For picture, see **central angle**.

The term arc is also used for a portion of any curve.

(See also **arc length; spherical trigonometry**.)

ARC LENGTH The length of an arc of a curve can be found with integration. Let ds represent a very small segment of the arc, and let dx and dy represent the x and y components of the arc. (See figure 9.)

Then:

$$ds = \sqrt{dx^2 + dy^2}$$

Rewrite this as:

$$ds = \sqrt{1 + \left(\frac{dy}{dx}\right)^2}\,dx$$

Now, suppose we need to know the length of the arc between the lines $x = a$ and $x = b$. Set up this integral:

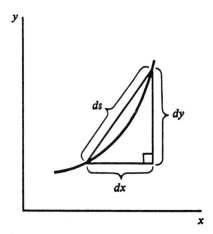

Figure 9 Arc length

$$s = \int_a^b \sqrt{1 + \left(\frac{dy}{dx}\right)^2}\, dx$$

For example, the length of the curve $y = x^{1.5}$ from a to b is given by the integral:

$$
\begin{aligned}
s &= \int_a^b \sqrt{1 + (1.5x^{.5})^2}\, dx \\
  &= \int_a^b \sqrt{1 + 2.25x}\, dx
\end{aligned}
$$

Let $u = 1 + 2.25x$; $dx = du/2.25$

$$s = \int_{1+2.25a}^{1+2.25b} (\sqrt{u}/2.25)\, du$$

$$= \frac{1}{1.5 \times 2.25} u^{1.5} \Big|_{1+2.25a}^{1+2.25b}$$

$$= \frac{(1 + 2.25b)^{1.5} - (1 + 2.25a)^{1.5}}{3.375}$$

ARCCOS If $x = \cos y$, then $y = \arccos x$. (See **inverse trigonometric functions**.)

ARCCSC If $x = \csc y$, then $y = \text{arccsc } x$. (See **inverse trigonometric functions**.)

ARCCTN If $x = \text{ctn } y$, then $y = \text{arcctn } x$. (See **inverse trigonometric functions**.)

ARCHIMEDES Archimedes (c 290 BC to c 211 BC) studied at Alexandria and then lived in Syracuse. He wrote extensively on mathematics and developed formulas for the volume and surface area of a sphere, and a way to calculate the circumference of a circle. He also developed the principle of floating bodies and invented military devices that delayed the capture of the city by the Romans.

ARCSEC If $x = \sec y$, then $y = \text{arcsec } x$. (See **inverse trigonometric functions**.)

ARCSIN If $x = \sin y$, then $y = \arcsin x$. (See **inverse trigonometric functions**.)

ARCTAN If $x = \tan y$, then $y = \arctan x$. (See **inverse trigonometric functions**.)

AREA The area of a two-dimensional figure measures how much of a plane it fills up. The area of a square of side a is defined as a^2. The area of every other plane figure is defined so as to be consistent with this definition. The area postulate in geometry says that if two figures are congruent, they have the same area. Area is measured in square units, such as square meters or square miles. The areas of some common figures are as follows:

Figure	Area
rectangle (sides a and b)	ab
parallelogram	(base) $\times$ (altitude)
triangle	$\frac{1}{2}$(base) $\times$ (altitude)
circle (radius r)	πr^2
ellipse (semimajor axis a, semiminor axis b)	$\pi a b$

The area of any polygon can be found by breaking the polygon up into many triangles. The areas of curved figures can often be found by the process of integration. (See **calculus**.)

ARGUMENT (1) The argument of a function is the independent variable that is put into the function. In the expression $\sin x$, x is the argument of the sine function.

(2) In logic an argument is a sequence of sentences (called premises) that lead to a resulting sentence (called the conclusion). (See **logic**.)

ARISTOTLE Aristotle (384 BC to 322 BC) wrote about many areas of human knowledge, including the field of logic. He was a student of Plato and also a tutor to Alexander the Great.

ARITHMETIC MEAN The arithmetic mean of a group of n numbers $(a_1, a_2, \ldots a_n)$, written as $\bar{a}$, is the sum of the numbers divided by n:

$$\bar{a} = \frac{a_1 + a_2 + a_3 + \ldots + a_n}{n}$$

The arithmetic mean is commonly called the average. For example, if your grocery bills for 4 weeks are \$10, \$15, \$12, and \$39, then the average grocery bill is $76/4 = \$19$.

ARITHMETIC PROGRESSION See **arithmetic sequence**.

ARITHMETIC SEQUENCE An arithmetic sequence is a sequence of n numbers of the form

$$a, a + b, a + 2b, a + 3b, \ldots, a + (n-1)b$$

ARITHMETIC SERIES An arithmetic series is a sum of an arithmetic sequence:

$$S = a + (a+b) + (a+2b) + (a+3b) + \ldots + [a + (n-1)b]$$

In an arithmetic series the difference between any two successive terms is a constant (in this case b). The sum of the first n terms in the arithmetic series above is

$$\sum_{i=0}^{n-1} (a + ib) = \frac{n}{2}[2a + (n-1)b]$$

For example:

$$3 + 5 + 7 + 9 + 11 + 13 = \frac{6}{2}[2(3) + (5)(2)] = 48$$

ASSOCIATIVE PROPERTY An operation obeys the associative property if the grouping of the numbers involved does not matter. Formally, the associative property of addition says that

$$(a + b) + c = a + (b + c)$$

for all a, b, and c.

The associative property for multiplication says that

$$(a \times b) \times c = a \times (b \times c)$$

For example:

$$(3 + 4) + 5 = 7 + 5 = 12 = 3 + (4 + 5) = 3 + 9$$
$$(5 \times 6) \times 7 = 30 \times 7 = 210 = 5 \times (6 \times 7) = 5 \times 42$$

ASYMPTOTE An asymptote is a straight line that is a close approximation to a particular curve as the curve goes off to infinity in one direction. The curve becomes very, very close to the asymptote line, but never touches it. For example, as x approaches infinity, the curve $y = 2^{-x}$ approaches very close to the line $y = 0$, but it never touches that line. See figure 10. (This is known as a horizontal asymptote.) As x approaches 3, the curve $y = 1/(x - 3)$ approaches the line $x = 3$. (This is known as a vertical asymptote.) For another example of an asymptote, see **hyperbola**.

AVERAGE The average of a group of numbers is the same as the **arithmetic mean**.

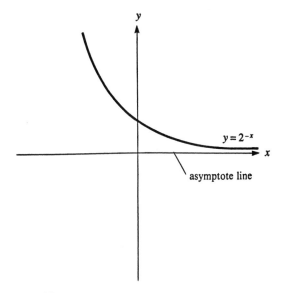

Figure 10

AXIOM An axiom is a statement that is assumed to be true without proof. Axiom is a synonym for postulate.

AXIS (1) The x-axis in Cartesian coordinates is the line $y = 0$. The y-axis is the line $x = 0$.

(2) The axis of a figure is a line about which the figure is symmetric. For example, the parabola $y = x^2$ is symmetric about the line $x = 0$. (See **axis of symmetry**.)

AXIS OF SYMMETRY An axis of symmetry is a line that passes through a figure in such a way that the part of the figure on one side of the line is the mirror image of the part of the figure on the other side of the line. (See **reflection**.) For example, an ellipse has two axes of symmetry: the major axis and the minor axis. (See **ellipse**.)

B

BASE (1) In the equation $x = \log_a y$, the quantity a is called the base. (See **logarithm**.)

(2) The base of a positional number system is the number of digits it contains. Our number system is a decimal, or base 10, system; in other words, there are 10 possible digits: 0, 1, 2, 3, 4, 5, 6, 7, 8, 9. For example, the number 123.789 means

$$1 \times 10^2 + 2 \times 10^1 + 3 \times 10^0 + 7 \times 10^{-1} + 8 \times 10^{-2} + 9 \times 10^{-3}$$

In general, if b is the base of a number system, and the digits of the number x are $d_4 d_3 d_2 d_1 d_0$ then $x = d_4 b^4 + d_3 b^3 + d_2 b^2 + d_1 b + d_0$

Computers commonly use base-2 numbers. (See **binary numbers**.)

(3) The base of a polygon is one of the sides of the polygon. For an example, see **triangle**. The base of a solid figure is one of the faces. For examples, see **cone, cylinder, prism, pyramid**.

BASIC FEASIBLE SOLUTION A basic feasible solution for a linear programming problem is a solution that satisfies the constraints of the problem where the number of nonzero variables equals the number of constraints. (By assumption we are ruling out the special case where more than two constraints intersect at one point, in which case there could be fewer nonzero variables than indicated above.)

Consider a linear programming problem with m constraints and n total variables (including slack variables). (See **linear programming**.) Then a basic feasible solution is a solution that satisfies the constraints of the problem and has exactly m nonzero variables and $n - m$ variables equal to zero. The basic feasible solutions will be at the corners of the feasible region, and an important theorem of linear programming states that, if there is an optimal solution, it will be a basic feasible solution.

BAYES Thomas Bayes (1702 to 1761) was an English mathematician who studied probability and statistical inference. (See **Bayes's rule**.)

BAYES'S RULE Bayes's rule tells how to find the conditional probability $Pr(B|A)$ (that is, the probability that event B will occur, given that event A has occurred), provided that $Pr(A|B)$ and $Pr(A|B^c)$ are known. (See **conditional probability**.) (B^c represents the event B-complement, which is the event that B will not occur.) Bayes's rule states:

$$Pr(B|A) = \frac{Pr(A|B)Pr(B)}{Pr(A|B)Pr(B) + Pr(A|B^c)Pr(B^c)}$$

For example, suppose that two dice are rolled. Let A be the event of rolling doubles, and let B be the event where the sum of the numbers on the two dice is greater than or equal to 8. Then

$$Pr(A) = \frac{6}{36} = \frac{1}{6}; \; Pr(B) = \frac{15}{36} = \frac{5}{12}$$

$$Pr(B^c) = \frac{21}{36} = \frac{7}{12}$$

$Pr(A|B)$ refers to the probability of obtaining doubles if the sum of the two numbers is greater than or equal to 8; this probability is $3/15 = 1/5$. There are 15 possible outcomes where the sum of the two numbers is greater than or equal to 8, and three of these are doubles: (4, 4), (5, 5), and (6, 6). Also, $Pr(A|B^c) = 3/21 = 1/7$ (the probability of obtaining doubles if the sum on the dice is less than 8). Then we can use Bayes's rule to find the probability that the sum of the two numbers will be greater than or equal to 8, given that doubles were obtained:

$$Pr(B|A) = \frac{\frac{1}{5} \times \frac{5}{12}}{\frac{1}{5} \times \frac{5}{12} + \frac{1}{7} \times \frac{7}{12}} = \frac{\frac{1}{12}}{\frac{1}{12} + \frac{1}{12}} = \frac{1}{2}$$

BERNOULLI Jakob Bernoulli (1655 to 1705) was a Swiss mathematician who studied concepts in what is now the calculus of variations, particularly the catenary curve. His brother Johann Bernoulli (1667 to 1748) also was a mathematician investigating these issues. Daniel Bernoulli (1700 to 1782, son of Johann) investigated mathematics and other areas. He developed Bernoulli's theorem in fluid mechanics, which governs the design of airplane wings.

BETWEEN In geometry point B is defined to be between points A and C if $AB + BC = AC$, where AB is the distance from point A to point B, and so on. This formal definition matches our intuitive idea that a point is between two points if it lies on the line connecting these two points and has one of the two points on each side of it.

BICONDITIONAL STATEMENT A biconditional statement is a compound statement that says one sentence is true if and only if the other sentence is true. Symbolically, this is written as $p \leftrightarrow q$, which means "$p \rightarrow q$" and "$q \rightarrow p$." (See **conditional statement**.) For example, "A triangle has three equal sides if and only if it has three equal angles" is a biconditional statement.

BINARY NUMBERS Binary (base-2) numbers are written in a positional system that uses only two digits: 0 and 1. Each digit of a binary number represents a power of 2. The rightmost digit is the 1's digit, the next digit to the left is the 2's digit, and so on.

Decimal	Binary
$2^0 = 1$	1
$2^1 = 2$	10
$2^2 = 4$	100
$2^3 = 8$	1000
$2^4 = 16$	10000

For example, the binary number 10101 represents

$$1 \times 2^4 + 0 \times 2^3 + 1 \times 2^2 + 0 \times 2^1 + 1 \times 2^0$$
$$= 16 + 0 + 4 + 0 + 1 = 21$$

Here is a table showing some numbers in both binary and decimal form:

Decimal	Binary	Decimal	Binary
0	0	11	1011
1	1	12	1100
2	10	13	1101
3	11	14	1110
4	100	15	1111
5	101	16	10000
6	110	17	10001
7	111	18	10010
8	1000	19	10011
9	1001	20	10100
10	1010	21	10101

Binary numbers are well suited for use by computers, since many electrical devices have two distinct states: on and off.

BINOMIAL A binomial is the sum of two terms. For example, $(ax + b)$ is a binomial.

BINOMIAL DISTRIBUTION Suppose that you conduct an experiment n times, with a probability of success of p each time. If X is the number of successes that occur in those n trials, then X will have the binomial distribution with parameters n and p. X is a discrete random variable whose probability function is given by

$$f(i) = Pr(X = i) = \binom{n}{i} p^i (1 - p)^{n-i}$$

In this formula $\binom{n}{i} = n!/[(n - i)!i!]$. (See **binomial theorem; factorial; combinations.**)

The expectation is $E(X) = np$; the variance is $\text{Var}(X) = np(1 - p)$. For example, roll a set of two dice five times, and let $X =$ the number of sevens that appear. Call it a "success" if a seven appears. Then the probability of success is $1/6$, so X has the binomial distribution with parameters $n = 5$ and $p = 1/6$. If you calculate the probabilities:

$$
\begin{aligned}
Pr(X = i) &= \frac{5!}{(5 - i)!i!} \left(\frac{1}{6}\right)^i \left(\frac{5}{6}\right)^{n-i} \\
Pr(X = 0) &= .402 \\
Pr(X = 1) &= .402 \\
Pr(X = 2) &= .161 \\
Pr(X = 3) &= .032 \\
Pr(X = 4) &= .003 \\
Pr(X = 5) &= .0001
\end{aligned}
$$

Also, if you toss a coin n times, and X is the number of heads that appear, then X has the binomial distribution with $p = \frac{1}{2}$:

$$Pr(X = i) = \binom{n}{i} 2^{-n}$$

BINOMIAL THEOREM The binomial theorem tells how to expand the expression $(a + b)^n$. Some examples of the powers of binomials are as follows:

$$
\begin{aligned}
(a + b)^0 &= 1 \\
(a + b)^1 &= a + b \\
(a + b)^2 &= a^2 + 2ab + b^2 \\
(a + b)^3 &= a^3 + 3a^2b + 3ab^2 + b^3 \\
(a + b)^4 &= a^4 + 4a^3b + 6a^2b^2 + 4ab^3 + b^4 \\
(a + b)^5 &= a^5 + 5a^4b + 10a^3b^2 + 10a^2b^3 + 5ab^4 + b^5
\end{aligned}
$$

Some patterns are apparent. The sum of the exponents for a and b is n in every term. The coefficients form an interesting pattern of numbers known as Pascal's triangle. This triangle is an array of numbers such that any entry is equal to the sum of the two entries above it.

In general, the binomial theorem states that

$$
(a + b)^n \simeq \binom{n}{0} a^n + \binom{n}{1} a^{n-1}b + \binom{n}{2} a^{n-2}b^2 \\
+ \cdots + \binom{n}{n-1} ab^{n-1} + \binom{n}{n} b^n
$$

The expression $\binom{n}{j}$ is called the binomial coefficient. It is defined to be

$$\binom{n}{j} = \frac{n!}{(n-j)!j!}$$

which is the number of ways of selecting n things, taken j at a time, if you don't care about the order in which the objects are selected. (See **combinations; factorial**.) For example:

$$\binom{n}{0} = \frac{n!}{n!0!} = 1$$

$$\binom{n}{1} = \frac{n!}{(n-1)!1!} = n$$

$$\binom{n}{2} = \frac{n!}{(n-2)!2!} = \frac{n(n-1)}{2}$$

$$\binom{n}{n-1} = \frac{n!}{1!(n-1)!} = n$$

$$\binom{n}{n} = \frac{n!}{0!n!} = 1$$

The binomial theorem can be proven by using **mathematical induction**.

BISECT To bisect means to cut something in half. For example, the perpendicular bisector of a line segment $\overline{AB}$ is the line perpendicular to the segment and halfway between A and B.

BOLYAI Janos Bolyai (1802 to 1860) was a Hungarian mathematician who developed a version of non-Euclidian geometry.

BOOLE George Boole (1815 to 1865) was an English mathematician who developed the symbolic analysis of logic now known as Boolean algebra, which is used in the design of digital computers.

BOOLEAN ALGEBRA Boolean algebra is the study of operations carried out on variables that can have only two values: 1 (true) or 0 (false). Boolean algebra was developed by George Boole in the 1850s; it is an important part of the theory of **logic** and has become of tremendous importance since the development of computers. Computers consist of electronic circuits (called flip-flops) that can be in either of two states, on or off, called 1 or 0. They are connected by circuits (called gates) that represent the logical operations of NOT, AND, and OR.

Here are some rules from Boolean algebra. In the following statements, p, q, and r represent Boolean variables and $\leftrightarrow$ represents "is equivalent to." Parentheses are used as they are in arithmetic: an operation inside parentheses is to be done before the operation outside the parentheses.

Double Negation:
$p \leftrightarrow$ NOT (NOT p)

Commutative Principle:
$(p$ AND $q) \leftrightarrow (q$ AND $p)$
$(p$ OR $q) \leftrightarrow (q$ OR $p)$

Associative Principle:
p AND $(q$ AND $r) \leftrightarrow (p$ AND $q)$ AND r
p OR $(q$ OR $r) \leftrightarrow (p$ OR $q)$ OR r

Distribution:
p AND $(q$ OR $r) \leftrightarrow (p$ AND $q)$ OR $(p$ AND $r)$
p OR $(q$ AND $r) \leftrightarrow (p$ OR $q)$ AND $(p$ OR $r)$

De Morgan's Laws:
(NOT p) AND (NOT q) $\leftrightarrow$ NOT (p OR q)
(NOT p) OR (NOT q) $\leftrightarrow$ NOT (p AND q)

Truth tables are a valuable tool for studying Boolean expressions. (See **truth table**.) For example, the first distributive property can be demonstrated with a truth table:

p	q	r	q OR r	p AND (q OR r)	p AND q	p AND r	(p AND q) OR (p AND r)
T	T	T	T	T	T	T	T
T	T	F	T	T	T	F	T
T	F	T	T	T	F	T	T
T	F	F	F	F	F	F	F
F	T	T	T	F	F	F	F
F	T	F	T	F	F	F	F
F	F	T	T	F	F	F	F
F	F	F	F	F	F	F	F

The fifth column and the last column are identical, so the sentence "p AND (q OR r)" is equivalent to the sentence "(p AND q) OR (p AND r)."

C

CALCULUS Calculus is divided into two general areas: differential calculus and integral calculus. The basic problem in differential calculus is to find the rate of change of a function. Geometrically, this means finding the slope of the tangent line to a function at a particular point; physically, this means finding the speed of an object if you are given its position as a function of time. The slope of the tangent line to the curve $y = f(x)$ at a point $(x, f(x))$ is called the *derivative*, written as y' or dy/dx, which can be found from this formula:

$$y' = \frac{dy}{dx} = \lim_{\Delta x \to 0} \frac{f(x + \Delta x) - f(x)}{\Delta x}$$

where "lim" is an abbreviation for "limit," and Δx means "change in x."

See **derivative** for a table of the derivatives of different functions. The process of finding the derivative of a function is called *differentiation*.

If f is a function of more than one variable, as in $f(x, y)$ then the partial derivative of f with respect to x (written as $\partial f/\partial x$) is found by taking the derivative of f with respect to x, while assuming that y remains constant. (See **partial derivative**.)

The reverse process of differentiation is integration (or antidifferentiation). Integration is represented by the symbol $\int$:

If $dy/dx = f(x)$, then:

$$y = \int f(x)dx = F(x) + C$$

This expression (called an indefinite integral) means that $F(x)$ is a function such that its derivative is equal to $f(x)$:

$$\frac{dF(x)}{dx} = f(x)$$

C can be any constant number; it is called the arbitrary constant of integration. A specific value can be assigned to C if an initial condition is known. (See **indefinite integral**.) See **integral** to learn procedures for finding integrals. Table 9 at the back of the book gives a table of some integrals.

A related problem is, What is the area under the curve $y = f(x)$ from $x = a$ to $x = b$? (Assume that $f(x)$ is continuous and always positive when $a < x < b$.) It turns out that this problem can be solved by integration:

$$(area) = F(b) - F(a)$$

where $F(x)$ is an antiderivative function: $dF(x)/dx = f(x)$. This area can also be written as a definite integral:

$$(area) = \int_a^b f(x)dx = F(b) - F(a)$$

(See **definite integral**.) In general:

$$\lim_{\Delta x \to 0, n \to \infty} \sum_{i=1}^{n} f(x_i)\Delta x = \int_a^b f(x)dx$$

where $\Delta x = \frac{b-a}{n}, x_1 = a, x_n = b$.

For other applications, see **arc length; surface area, figure of revolution; volume, figure of revolution; centroid**.

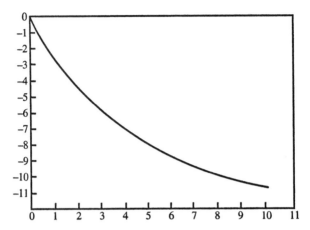

Figure 11 Cycloid curve: the fastest way for a ball to reach the end

CALCULUS OF VARIATIONS In calculus of variations, the problem is to determine a curve $y(x)$ that minimizes (or maximizes) the integral of a specified function over a specific range:

$$\int_a^b f(x, y, y') dx$$

where y' is the derivative of y with respect to x (also known as dy/dx). The optimal curve y will solve this differential equation:

$$\frac{\partial f}{\partial y} - \frac{d}{dx}\left(\frac{\partial f}{\partial y'}\right) = 0$$

Here is an example of this type of problem. You need to design a ramp that will allow a ball to roll downhill between point $(0,0)$ and point $(10,-10)$ in the least possible time. The correct answer is not a straight line. Instead, the ramp should slope down-

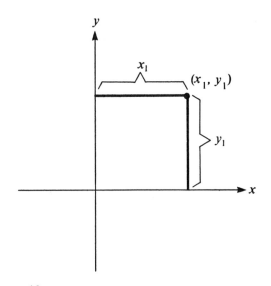

Figure 12   Cartesian coordinates

ward steeply at the beginning so the ball picks up speed more quickly. The solution to this problem turns out to be the **cycloid** curve:

$$x = a(\theta - \sin\theta) \quad y = -a(1 - \cos\theta)$$

where the value of a is adjusted so the curve passes through the desired final point (in our case, a equals 5.729). (See figure 11.)

CARTESIAN COORDINATES A Cartesian coordinate system is a system whereby points on a plane are identified by an ordered pair of numbers, representing the distances to two perpendicular axes. The horizontal axis is usually called the x-axis, and the vertical axis is usually called the y-axis. (See figure 12). The x-coordinate is always listed first in an ordered pair such as (x_1, y_1). Cartesian coordinates are also called rectangular coordinates to distinguish them from polar coordinates. A three-dimensional Cartesian coordinate system can be constructed by drawing a z-axis perpendicular to the x- and y-axes. A three-dimensional coordinate system can label any point in space.

CARTESIAN PRODUCT The Cartesian product of two sets, A and B (written $A \times B$), is the set of all possible ordered pairs that have a member of A as the first entry and a member of B as the second entry. For example, if $A = (x, y, z)$ and $B = (1, 2)$, then $A \times B = \{(x, 1), (x, 2), (y, 1), (y, 2), (z, 1), (z, 2)\}$.

CATENARY A catenary is a curve represented by the formula

$$y = \frac{1}{2}a(e^{x/a} + e^{-x/a})$$

The value of e is about 2.718. (See **e.**) The value of a is the y intercept. The catenary can also be represented by the hyperbolic cosine function $y = \cosh x$

The curve formed by a flexible rope allowed to hang between two posts will be a catenary. (See figure 13.)

CENTER (1) The center of a circle is the point that is the same distance from all of the points on the circle.

(2) The center of a sphere is the point that is the

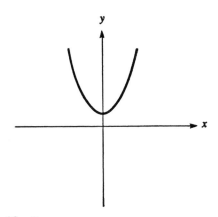

Figure 13 Catenary

same distance from all of the points on the sphere.

(3) The center of an ellipse is the point where the two axes of symmetry (the major axis and the minor axis) intersect.

(4) The center of a regular polygon is the center of the circle that can be inscribed in that polygon.

CENTER OF MASS See **centroid.**

CENTRAL ANGLE A central angle is an angle that has its vertex at the center of a circle. (See figure 14.)

CENTRAL LIMIT THEOREM See **normal distribution.**

CENTROID The centroid is the center of mass of an object. It is the point where the object would balance if supported by a single support. For a triangle, the centroid is the point where the three medians in-

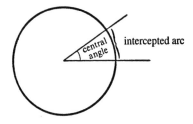

Figure 14

tersect. For a one-dimensional object of length L, the centroid can be found by using the integral

$$\frac{\int_0^L x\rho dx}{\int_0^L \rho dx}$$

where $\rho(x)$ represents the mass per unit length of the object at a particular location x. The centroid for two- or three-dimensional objects can be found with double or triple integrals.

CHAIN RULE The chain rule in calculus tells how to find the derivative of a composite function. If f and g are functions, and if $y = f(g(x))$, then the chain rule states that

$$\frac{dy}{dx} = \frac{df}{dg}\frac{dg}{dx}$$

For example, suppose that $y = \sqrt{1 + 3x^2}$ and you are required to define these two functions:

$$g(x) = 1 + 3x^2; \quad f(g) = \sqrt{g}$$

Then y is a composite function: $y = f(g(x))$, and

$$\frac{df}{dg} = \frac{1}{2}g^{-1/2}$$

$$\frac{dg}{dx} = 6x$$

$$\frac{dy}{dx} = \frac{1}{2}g^{-1/2}6x = 3x(1 + 3x^2)^{-1/2}$$

Here are other examples (assume that a and b are constants):

$y = \sin(ax + b)$	$\frac{dy}{dx} = a\cos(ax + b)$
$y = \ln(ax + b)$	$\frac{dy}{dx} = \frac{a}{ax+b}$
$y = e^{ax}$	$\frac{dy}{dx} = ae^{ax}$

CHAOS Chaos is the study of systems with the property that a small change in the initial conditions can lead to very large changes in the subsequent evolution of the system. Chaotic systems are inherently unpredictable. The weather is an example; small changes in the temperature and pressure over the ocean can lead to large variations in the future development of a storm system. However, chaotic systems can exhibit certain kinds of regularities.

CHARACTERISTIC The characteristic is the integer part of a common logarithm. For example, log $115 = 2.0607$, where 2 is the characteristic and .0607 is the mantissa.

CHEBYSHEV Pafnuty Lvovich Chebyshev (1821 to 1894) was a Russian mathematician who studied probability, among other areas of mathematics. (See **Chebyshev's theorem**.)

CHEBYSHEV'S THEOREM Chebyshev's theorem states that, for any group of numbers, the fraction that will be within k standard deviations of the mean will be at least $1 - 1/k^2$. For example, if $k = 2$, the formula gives the value of $1 - \frac{1}{4} = \frac{3}{4}$. Therefore, for any group of numbers at least 75 percent of them will be within two standard deviations of the mean.

CHI SQUARE DISTRIBUTION If $Z_1, Z_2, Z_3, \ldots,$ Z_n are independent and identically distributed standard normal random variables, then the random variable

$$S = Z_1^2 + Z_2^2 + Z_3^2 + \ldots + Z_n^2$$

will have the chi square distribution with n degrees of freedom. The chi-square distribution with n degrees of freedom is symbolized by χ_n^2, since χ is the Greek letter chi. For the χ_n^2 distribution, $E(X) = n$ and $Var(X) = 2n$.

The chi-square distribution is used extensively in statistical estimation. (See **chi-square test.**) It is also used in the definition of the t-distribution.

Table 5 lists some values for the chi-square cumulative distribution function.

CHI-SQUARE TEST The chi-square test provides a method for testing whether a particular probability distribution fits an observed pattern of data, or for testing whether two factors are independent. The chi-square test statistic is calculated from this formula:

$$\frac{(f_1 - f_1*)^2}{f_1*} + \frac{(f_2 - f_2*)^2}{f_2*} + \cdots + \frac{(f_n - f_n*)^2}{f_n*}$$

where f_i is the actual frequency of observations, and f_i* is the expected frequency of observations if the null hypothesis is true, and n is the number of comparisons being made. If the null hypothesis is true,

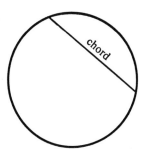

Figure 15

then the test statistic will have a chi-square distribution. The number of degrees of fi dom depends on the number of observations. If the computed value of the test statistic is too large, the null hypothesis is rejected. (See **hypothesis testing**.)

CHORD A chord is a line segment that connects two points on a curve. (See figure 15.)

CIRCLE A circle is the set of points in a plane that are all a fixed distance from a given point. The given point is known as the center. The distance from the center to a point on the circle is called the radius (symbolized by r). The diameter is the farthest distance across the circle; it is equal to twice the radius. The circumference is the distance you would have to walk if you walked all the way around the circle. The circumference equals $2\pi r$, where $\pi = 3.14159...$ (See **pi**.)

The equation for a circle with center at the origin is $x^2 + y^2 = r^2$. This equation is derived from the distance formula. If the center is at (h, k), the

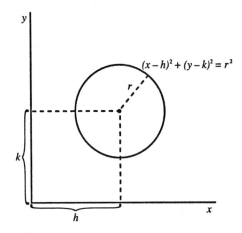

Figure 16 Circle

equation is

$$(x - h)^2 + (y - k)^2 = r^2$$

(See figure 16.)

The area of a circle equals πr^2. To show this, imagine dividing the circle into n triangular sectors, each with an area approximately equal to $\frac{rC}{2n}$. (See figure 17.) To get the total area of the circle, multiply by n:

$$(area) = \frac{rC}{2} = \pi r^2$$

(To be exact, you have to take the limit as the number of triangles approaches infinity.)

CIRCULAR FUNCTIONS The circular functions are the same as the trigonometric functions.

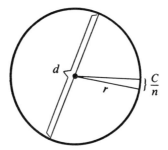

Figure 17

CIRCUMCENTER The circumcenter of a triangle is the center of the circle that can be circumscribed about the triangle. It is at the point where the perpendicular bisectors of the three sides cross. (See **triangle.**)

CIRCUMCIRCLE The circumcircle for a triangle is the circle that can be circumscribed about the triangle. The three vertices of the triangle are points on the circle. For illustration, see **triangle.**

CIRCUMFERENCE The circumference of a closed curve (such as a circle) is the total distance around the curve. The circumference of a circle is $2\pi r$, where r is the radius. (See **pi.**) Formally, the circumference of a circle is defined as the limit of the perimeter of a regular inscribed n-sided polygon as the number of sides goes to infinity.

CIRCUMSCRIBED A circumscribed circle is a circle that passes through all of the vertices of a polygon. For an example, see **triangle.** For contrast,

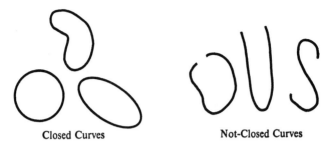

Closed Curves   Not-Closed Curves

Figure 18

see **inscribed**. In general, a figure is circumscribed about another if it surrounds it, touching it at as many points as possible.

CLOSED CURVE A closed curve is a curve that completely encloses an area. (See figure 18.)

CLOSED INTERVAL A closed interval is an interval that contains its endpoints. For example, the interval $0 \leq x \leq 1$ is a closed interval because the two endpoints (0 and 1) are included. For contrast, see **open interval**.

CLOSED SURFACE A closed surface is a surface that completely encloses a volume of space. For example, a sphere (like a basketball) is a closed surface, but a teacup is not.

CLOSURE PROPERTY An arithmetic operation obeys the closure property with respect to a given set of numbers if the result of the operation will always be in that set if the operands (the input numbers)

are. For example, the operation of addition is closed
with respect to the integers, but the operation of
division is not. (If a and b are integers, $a + b$ will
always be an integer, but a/b may or may not be.)

Operation	Natural Numbers	Integers	Rational Numbers	Real Numbers
addition	closed	closed	closed	closed
subtraction	not closed	closed	closed	closed
division	not closed	not closed	closed	closed
root extraction	not closed	not closed	not closed	not closed

COEFFICIENT Coefficient is a technical term for
something that multiplies something else (usually applied to a constant multiplying a variable). In the
quadratic equation

$$Ax^2 + Bxy + Cy^2 + Dx + Ey + F = 0$$

A is the coefficient of x^2, B is the coefficient of xy,
and so on.

COEFFICIENT OF DETERMINATION The
coefficient of determination is a value between 0 and
1 that indicates how well the variations in the independent variables in a regression explain the variations in the dependent variable. It is symbolized by
r^2. (See **regression; multiple regression.**)

COEFFICIENT OF VARIATION The coefficient
of variation for a list of numbers is equal to the
standard deviation for those numbers divided by the
mean. It indicates how big the dispersion is in comparison to the mean.

COFUNCTION Each trigonometric function has a cofunction. Cosine is the cofunction for sine, cotangent is the cofunction for tangent, and cosecant is the cofunction for secant. The cofunction of a trigonometric function $f(x)$ is equal to $f(\pi/2 - x)$. The name cofunction is used because $\pi/2 - x$ is the complement of x. For example, $\cos(x) = \sin(\pi/2 - x)$.

COLLINEAR A set of points is collinear if they all lie on the same line. (Note that any two points are always collinear.)

COMBINATIONS The term combinations refers to the number of possible ways of arranging objects chosen from a total sample of size n if you don't care about the order in which the objects are arranged. The number of combinations of n things, taken j at a time, is $n!/[(n - j)!j!]$, which is written as

$$\binom{n}{j} = \frac{n!}{(n - j)!j!}$$

(See **factorial; binomial theorem**.)

For example, the number of possible poker hands is equal to the number of possible combinations of five objects drawn (without replacement) from a sample of 52 cards. The number of possible hands is therefore:

$$\binom{52}{5} = \frac{52!}{47!5!} = \frac{52 \times 51 \times 50 \times 49 \times 48}{5 \times 4 \times 3 \times 2 \times 1}$$
$$= 2,598,960$$

This formula comes from the fact that there are n ways to choose the first object, $n - 1$ ways to choose the second object, and therefore

$$n \times (n - 1) \times (n - 2) \times ...(n - j + 2) \times (n - j + 1)$$

ways of choosing all j objects. This expression is equal to $n!/(n-j)!$. However, this method counts each possible ordering of the objects separately. (See **permutations**.) Many times the order in which the objects are chosen doesn't matter. To find the number of combinations, we need to divide by $j!$, which is the total number of ways of ordering the j objects. That makes the final result for the number of combinations equal to $n!/[(n-j)!j!]$.

Counting the number of possible combinations for arranging a group of objects is important in probability. Suppose that both you and your dream lover (whom you're desperately hoping to meet) are in a class of 20 people, and five people are to be randomly selected to be on a committee. What is the probability that both you and your dream lover will be on the committee? The total number of ways of choosing the committee is

$$\binom{20}{5} = \frac{20!}{5!15!} = 15,504$$

Next, you need to calculate how many possibilities include both of you on the committee. If you've both been selected, then the other three members need to be chosen from the 18 remaining students, and there are

$$\binom{18}{3} = \frac{18!}{3!15!} = 816$$

ways of doing this. Therefore the probability that you'll both be selected is $816/15,504 = .053$. Your chances improve if the size of the committee increases. The table lists the probability of your both being selected if the committee contains s members:

s	Probability
2	.005
3	.016
4	.031
5	.053
6	.079
7	.111
8	.147
9	.189
10	.237
15	.553
18	.804
19	.900
20	1.000

COMMON LOGARITHM A common logarithm is a logarithm to the base 10. In other words, if $y = \log_{10} x$, then $x = 10^y$. Often $\log_{10} x$ is written as $\log x$, without the subscript 10. (See **logarithm**.) Here is a table of some common logarithms (expressed as four-digit decimal approximations):

x	$\log x$	x	$\log x$
1	0	7	0.8451
2	0.3010	8	0.9031
3	0.4771	9	0.9542
4	0.6021	10	1.0000
5	0.6990	50	1.6990
6	0.7782	100	2.0000

COMMUTATIVE PROPERTY An operation obeys the commutative property if the order of the

Figure 19 Compass

two numbers involved doesn't matter. The commutative property for addition states that

$$a + b = b + a$$

for all a and b. The commutative property for multiplication states that

$$ab = ba$$

for all a and b. For example, $3 + 6 = 6 + 3 = 9$, and
$6 \times 7 = 7 \times 6 = 42$. Neither subtraction, division, nor
exponentiation obeys the commutative property:

$$3 \neq 3 - 5,\ \frac{3}{4} \neq \frac{4}{3},\ 2^3 \neq 3^2$$

COMPASS A compass is a device consisting of two
   adjustable legs (figure 19), used for drawing circles
   and measuring off equal distance intervals. (See **geometric construction**.)

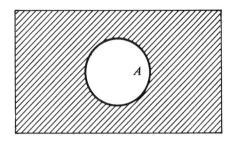

Figure 20 Complement of Set A

COMPLEMENT OF A SET The complement of a set A consists of the elements in a particular universal set that are not elements of set A. In the Venn diagram (figure 20) the shaded region is the complement of set A.

COMPLEMENTARY ANGLES Two angles are complementary if the sum of their measures is 90 degrees ($= \pi/2$ radians). For example, a 35° angle and a 55° angle are complementary. The two smallest angles in a right triangle are complementary.

COMPLETING THE SQUARE Sometimes an algebraic equation can be simplified by adding an expression to both sides that makes one part of the equation a perfect square.

For example, see **quadratic equation**.

COMPLEX FRACTION A complex fraction is a fraction in which either the numerator or the denominator or both contain fractions. For example,

$$\frac{\frac{2}{3}}{\frac{3}{4}}$$

is a complex fraction. To simplify the complex fraction, multiply both the numerator and the denominator by the reciprocal of the denominator:

$$\frac{\frac{2}{3}}{\frac{3}{4}} = \frac{\frac{2}{3} \times \frac{4}{3}}{\frac{3}{4} \times \frac{4}{3}} = \frac{\frac{8}{9}}{1} = \frac{8}{9}$$

COMPLEX NUMBER A complex number is formed by adding a pure imaginary number to a real number. The general form of a complex number is $a + bi$, where a and b are both real numbers and i is the imaginary unit: $i^2 = -1$. The number a is called the real part of the complex number, and b is the imaginary part. Two complex numbers are equal to each other only when both their real parts and their imaginary parts are equal to each other.

Complex numbers can be illustrated on a two-dimensional graph, much like a system of Cartesian coordinates. The real axis is the same as the real number line, and the imaginary axis is a line drawn perpendicular to the real axis. (See figure 21.)

To add two complex numbers. add the real parts and the imaginary parts separately:

$$(a + bi) + (c + di) = (a + c) + (b + d)i$$

Two complex numbers can be multiplied in the same way that you multiply two binomials:

$$\begin{aligned}(a + bi)(c + di) &= a(c + di) + bi(c + di)\\ &= ac + adi + bci + bdi^2\\ &= (ac - bd) + (ad + bc)i\end{aligned}$$

The absolute value of a complex number $(a + bi)$ is the distance from the point representing that number

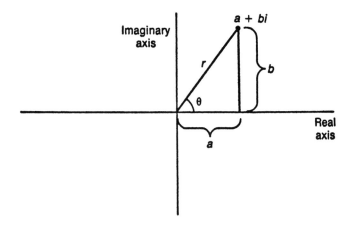

Figure 21 Complex Number

in the complex plane to the origin, which is equal
to $\sqrt{a^2 + b^2}$. The complex conjugate of $(a + bi)$ is
defined to be $(a - bi)$. The product of any complex
number with its conjugate will be a real number,
equal to the square of its absolute value:

$$(a + bi)(a - bi) = a^2 + abi - abi - b^2 i^2 = a^2 + b^2$$

Complex numbers are also different from real
numbers in that you can't put them in order.

Complex numbers can also be expressed in polar
form:

$$(a + bi) = r(\cos\theta + i\sin\theta)$$

where r is the absolute value ($r = \sqrt{a^2 + b^2}$) and θ
is the angle of inclination: $b/a = \tan\theta$. (See **polar
coordinates**.)

Multiplication is easy for two complex numbers
in polar form:

$$[r_1(\cos\theta_1 + i\sin\theta_1)] \times [r_2(\cos\theta_2 + i\sin\theta_2)]$$
$$= r_1 r_2[\cos(\theta_1 + \theta_2) + i\sin(\theta_1 + \theta_2)]$$

In words: to multiply two polar form complex numbers, multiply their absolute values and then add their angles.

To raise a polar form complex number to a power, use this formula:

$$[r(\cos\theta + i\sin\theta)]^n = r^n[\cos(n\theta) + i\sin(n\theta)]$$

(See also **De Moivre's theorem**.)

COMPONENT In the vector (a, b, c), the numbers a, b, and c are known as the components of the vector.

COMPOSITE FUNCTION A composite function is a function that consists of two functions arranged in such a way that the output of one function becomes the input of the other function. For example, if $f(u) = \sqrt{u} + 3$, and $g(x) = 5x$, then the composite function $f(g(x))$ is the function $\sqrt{5x} + 3$. To find the derivative of a composite function, see **chain rule**.

COMPOSITE NUMBER A composite number is a natural number that is not a prime number. Therefore, it can be expressed as the product of two natural numbers other than itself and 1.

COMPOUND INTEREST If A dollars are invested in an account paying compound interest at an annual rate r, then the balance in the account after n years will be $A(1 + r)^n$. The same formula works if the compounding period is different from one year,

Figure 22   Concave set

provided that n is the number of compounding periods and r is the rate per period. For example, the interest might be compounded once per month.

COMPOUND SENTENCE In logic, a compound sentence is formed by joining two or more simple sentences together with one or more connectives, such as AND, OR, NOT, or IF/THEN. (See **logic; Boolean algebra**.)

CONCAVE A set of points is concave if it is possible to draw a line segment that connects two points that are in the set, but includes also some points that are not in the set. (See figure 22.) Note that a concave figure looks as though it has "caved" in. For contrast, see **convex**.

In figure 23, curve A is oriented so that its concave side is down; curve B is oriented so that its concave side is up. If the curve represents the graph of $y = f(x)$, then the curve will be oriented concave up if the second derivative y'' is positive; it will be oriented concave down if the second derivative is negative.

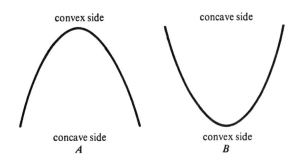

convex side

concave side

concave side
A

convex side
B

Figure 23

CONCLUSION The conclusion is the phrase in an argument that follows as a result of the premises. (See **logic**.) In a conditional statement the conclusion is the "then" part of the statement. It is the part that is true if the antecedent (the "if" part) is true. For example, in the statement "If he likes pizza, then he likes cheese," the conclusion is the clause "he likes cheese." The conclusion of a conditional statement is also called the consequent.

CONDITIONAL EQUATION A conditional equation is an equation that is true only for some values of the unknowns contained in the equation. For contrast, see **identity**.

CONDITIONAL PROBABILITY The conditional probability that event *A* will occur, given that event *B* has occurred, is written $\Pr(A|B)$ (read as "*A* given *B*"). It can be found from this formula:

$$\Pr(A|B) = \frac{\Pr(A \text{ AND } B)}{\Pr(B)}$$

For example, suppose you toss two dice. Let A be the event that the sum is 8; let B be the event that the number on the first die is 5. If you don't know the number of the first die, then you can find that $\Pr(A) = 5/36$. Using the conditional probablity formula, we can find:

$$\Pr(A|B) = \frac{1/36}{1/6} = \frac{6}{36} = \frac{1}{6}$$

Therefore, a knowledge of the number on the first die has changed the probability that the sum will be 8. If C is the event that the first die is 1, then $\Pr(A|C) = 0$. (See also **Bayes's rule.**)

CONDITIONAL STATEMENT A conditional statement is a statement of this form: "If a is true, then b is true." Symbolically, this is written as $a \rightarrow b$ ("a implies b"). For example, the statement "If a triangle has three equal sides, then it has three equal angles" is true, but the statement "If a quadrilateral has four equal sides, then it has four equal angles" is false.

CONE A cone (figure 24) is formed by the union of all line segments that connect a given point (called the vertex) and the points on a closed curve that is not in the same plane as the vertex. If the closed curve is a circle, then the cone is called a circular cone. The region enclosed by the circle is called the base. The distance from the plane containing the base to the vertex is called the altitude. The volume of the cone is equal to $\frac{1}{3}$(base area)(altitude).

Each line segment from the vertex to the circle is called an element of the cone. An ice cream cone is an example of a cone. The term cone also refers to the

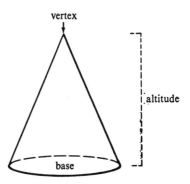

Figure 24 Cone

figure formed by all possible lines that pass through both the vertex point and a given circle. This type of cone goes off to infinity in two directions. (See **conic section**.)

CONFIDENCE INTERVAL A confidence interval is an interval based on observations of a sample constructed so that there is a specified probability that the interval contains the unknown true value of a population parameter. It is common to calculate confidence intervals that have a 95 percent probability of containing the true value.

    For example, suppose that you are trying to estimate the mean weight of loaves of bread produced at a bakery. It would be too expensive to weigh every single loaf, but you can estimate the mean by selecting and weighing a random sample of loaves. Suppose that the weights of the entire population of loaves have a normal distribution with mean μ, whose value is unknown, and a standard deviation sigma σ, whose value is known. Suppose also that you have selected a sample of n loaves and have found that the average weight of this sample is $\overline{x}$. (The bar over the x stands for "average.") Because of the properties of the normal distribution, $\overline{x}$ will have a normal distribution with mean μ and standard deviation $\sigma/\sqrt{n}$.

Now define Z as follows:

$$Z = \frac{\sqrt{n}(\overline{x} - \mu)}{\sigma}$$

Z will have a standard normal distribution (that is, a normal distribution with mean 0 and standard deviation 1). There is a 95 percent chance that a standard normal random variable will be between -1.96 and 1.96:

$$Pr(-1.96 < Z < 1.96) = .95$$

(See Table 4.) Therefore:

$$\Pr(-1.96 < \frac{\sqrt{n}(\overline{X} - \mu)}{\sigma} < 1.96) = .95.$$

which can be rewritten as

$$\Pr(\overline{x} - \frac{1.96\sigma}{\sqrt{n}} < \mu < \overline{x} + \frac{1.96\sigma}{\sqrt{n}}) = .95$$

The last equation tells you how to calculate the confidence interval. There is a 95 percent chance that the interval from $\overline{x} - 1.96\sigma/\sqrt{n}$ to $\overline{x} + 1.96\sigma/\sqrt{n}$ will contain the true value of the mean μ.

However, in many practical situations you will not know the true value of the population standard deviation, σ, and therefore cannot use the preceding method. Instead, after selecting your random sample of size n, you will need to calculate both the sample average, $\overline{x}$, and the sample standard deviation, s:

$$s = \sqrt{\frac{(x_1 - \overline{x})^2 + (x_2 - \overline{x})^2 + \ldots + (x_n - \overline{x})^2}{n-1}}$$

The confidence interval calculation is based on the fact that the quantity $T = \sqrt{n}(\overline{x} - \mu)/s$ will have a t-distribution with $n-1$ degrees of freedom. (See **t distribution**.) Note that the quantity T is the same as the quantity Z used above, except that the known value of the sample standard deviation s has been substituted for the population standard deviation, σ, which is now unknown. Now you need to look in a t-distribution table for a value (a) such that $\Pr(-a < T < a) = .95$, where T has a t distribution with the appropriate degrees of freedom. See Table 7 at the back of the book. Then the 95 percent confidence interval for the unknown value of μ is from

$$\overline{x} - \frac{as}{\sqrt{n}} \text{ to } \overline{x} + \frac{as}{\sqrt{n}}$$

For example, suppose you are investigating the mean commuting time along a particular route into the city. You have recorded the commuting times for 7 days:

$$39, 43, 29, 52, 35, 38, 39$$

and would like to calculate a 95 percent confidence interval for the mean commuting time. Calculate the sample average, $\overline{x} = 39.286$. Then calculate the sample standard deviation $s = 7.088$. Look in Table 7 for a t-distribution with $7 - 1 = 6$ degrees of freedom to find the value $a = 2.447$. Then the 95 percent confidence interval is

$$39.286 \pm \frac{2.447 \times 7.088}{\sqrt{7}}$$

which is from 32.730 to 45.841.

CONGRUENT Two polygons are congruent if they have exactly the same shape and exactly the same size. In other words, if you pick one of the polygons up and put it on top of the other, the two would match exactly. Each side of one polygon is exactly the same length as one side of the congruent polygon. These two sides with the same length are called corresponding sides. Also, each angle on one polygon has a corresponding angle on the other polygon. All of the pairs of corresponding angles are equal. See **triangle** for some examples of ways to prove that two triangles are congruent.

CONIC SECTIONS The four curves—circles, ellipses, parabolas, and hyperbolas (figures 25 and 26)— are called conic sections because they can be formed by the intersection of a plane with a right circular cone. If the plane is perpendicular to the axis of the cone, the intersection will be a circle. If the plane is slightly tilted, the result will be an ellipse. If the plane is parallel to one element of the cone, the result will be a parabola. If the plane intersects both parts of the cone, the result will be a hyperbola. (Note that a hyperbola has two branches.)

There is another definition of conic sections that makes it possible to define parabolas, ellipses, and hyperbolas by one equation. A conic section can be defined as a set of points such that the distance from a fixed point divided by the distance from a fixed line is a constant. The fixed point is called the focus, the fixed line is called the directrix, and the constant ratio is called the eccentricity of the conic section, or e. When $e = 1$ this definition exactly matches the definition of a parabola. If $e \neq 1$, you can find the

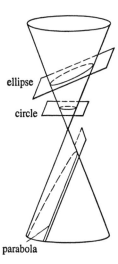

Figure 25 Conic Sections

equation for a conic section with the line $x = 0$ as the directrix and the point $(p, 0)$ as the focus (figure 27):

$$\frac{\sqrt{(x - p)^2 + y^2}}{x} = e$$

Simplifying:

$$x^2(1 - e^2) - 2px + y^2 + p^2 = 0$$

$$x^2 - \frac{2px}{1 - e^2} + \frac{y^2}{1 - e^2} + \frac{p^2}{1 - e^2} = 0$$

Complete the square by adding and subtracting $p^2/(1 - e^2)^2$:

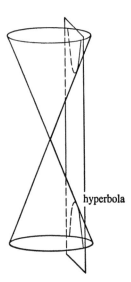

Figure 26 Hyperbola as a Conic Section

$$x^2 - \frac{2px}{1-e^2} + \frac{p^2}{(1-e^2)^2} - \frac{p^2}{(1-e^2)^2} + \frac{y^2}{1-e^2} + \frac{p^2}{1-e^2} = 0$$

$$\left[x - \left(\frac{p}{1-e^2}\right)\right]^2 + \frac{y^2}{1-e^2} = \frac{e^2p^2}{(1-e^2)^2}$$

This equation can be rewritten as

$$\frac{(x-h)^2}{a^2} + \frac{y^2}{B} = 1$$

where

$$h = \frac{p}{1-e^2}, \; a^2 = \frac{e^2p^2}{(1-e^2)^2}$$

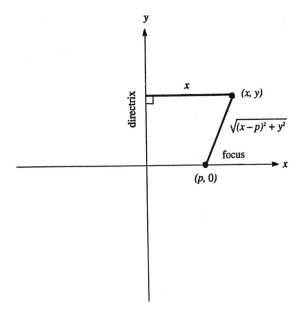

Figure 27 Definition of conic sections

and

$$B = \frac{e^2 p^2}{1 - e^2}$$

If $e < 1$, then B is positive, and this is the standard equation of an ellipse with center at $(h, 0)$. If $e > 1$, then B is negative, and this is the standard equation of a hyperbola.

CONJECTURE A conjecture is a statement that seems to be true, but it has not yet been proved. For an example, see **Fermat's last theorem**. For contrast, see **theorem**.

CONJUGATE The conjugate of a complex number is
formed by reversing the sign of the imaginary part.
The conjugate of $a + bi$ is $a - bi$. (See **complex
number**.) The product of a complex number with
its conjugate will always be a nonnegative real number:

$$(a + bi)(a - bi) = a^2 - abi + abi - b^2i^2 = a^2 + b^2$$

If a complex number $a + bi$ occurs in the denominator of a fraction, it helps to multiply both the
numerator and the denominator of the fraction by
$a - bi$:

$$\frac{3 + 2i}{4 + 6i} = \frac{(3 + 2i)(4 - 6i)}{(4 + 6i)(4 - 6i)} = \frac{12 - 18i + 8i - 12i^2}{16 - 24i + 24i - 36i^2}$$
$$- \frac{6}{13} - \frac{5}{26}i$$

CONJUNCTION A conjunction is an AND statement of this form: "A and B." It is true only if
both A and B are true. For example, the statement
"Two points determine a line and three noncollinear
points determine a plane" is true, but the statement
"Triangles have three sides and pentagons have four
sides" is false.

CONSEQUENT The consequent is the part of a conditional statement that is true if the other part (the
antecedent) is true. The consequent is the "then"
part of a conditional statement. For example, in the
statement "If he likes pizza, then he likes cheese," the
consequent is the clause "he likes cheese." The consequent is also called the conclusion of a conditional
statement.

Continuous   Discontinuous

Figure 28

CONSISTENT ESTIMATOR A consistent estimator is an estimator that tends to converge toward the
true value of the parameter it is trying to estimate
as the sample size becomes larger. (See **statistical
inference.**)

CONSTANT A constant represents a quantity that
does not change. It can be expressed either as a
numeral or as a letter (or other variable name) whose
value is taken to be a constant.

CONTINUOUS A continuous function is one that
you can graph without lifting your pencil from the
paper. (See figure 28.) Most functions that have
practical applications are continuous, but it is easy
to think of examples of discontinuous functions. The
formal definition of continuous is: The graph of $y =
f(x)$ is continuous at a point a if (1) $f(a)$ exists; (2)
$\lim_{x \to a} f(x)$ exists; and (3) $\lim_{x \to a} f(x) = f(a)$. A
function is continuous if it is continuous at each point
in its domain.

CONTINUOUS RANDOM VARIABLE A continuous random variable is a random variable that
can take on any real-number value within a certain

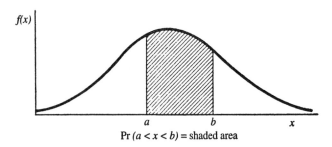

$\Pr(a < x < b) =$ shaded area

Figure 29 Density function for continuous random variable

range. It is characterized by a density function curve such that the area under the curve between two numbers represents the probability that the random variable will be between those two numbers. (See figure 29.)

The area can be expressed by this integral:

$$\Pr(a < X < b) = \int_a^b f(x)dx$$

where X is the random variable and $f(x)$ is the density function. The density function must satisfy

$$\int_{-\infty}^{\infty} f(x)dx = 1$$

In words: the total area under the density function must be one, or $\Pr(-\infty < X < \infty) = 1$

For examples of continuous random variable distributions, see **normal distribution, chi-square distribution, t-distribution, F-distribution.** For contrast, see **discrete random variable.**

CONTRADICTION A contradiction is a statement that is necessarily false because of its logical structure, regardless of the facts. For example, the statement "p AND (NOT p)" is false, regardless of what p represents. The negation of a contradiction is a tautology.

CONTRAPOSITIVE The contrapositive of the statement $A \rightarrow B$ is the statement (NOT B) $\rightarrow$ (NOT A). The contrapositive is equivalent to the original statement. If the original statement is true, the contrapositive is true; if the original statement is false, the contrapositive is false. For example, the statement "If x is a rational number, then x is a real number" has the contrapositive "If x is not a real number, then it is not a rational number."

CONVERGENT SERIES A convergent series is an infinite series that has a finite sum. For example, the series

$$1 + x + x^2 + x^3 + x^4 + \cdots$$

is convergent if $|x| < 1$ (in which case the sum of the series is

$$\frac{1}{1 - x}$$

If $|x| \geq 1$, then the sum of the series is infinite and it is called a **divergent series**.

CONVERSE The converse of an IF-THEN statement is formed by interchanging the "if" part and the "then" part:

statement: $a \rightarrow b$
converse: $b \rightarrow a$

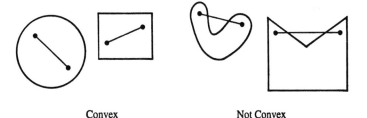

Convex   Not Convex

Figure 30

The converse of a true statement may be true, or it may be false. For example:

Statement (true) "If a triangle is a right triangle, then the square of the length of the longest side is equal to the sums of the squares of the lengths of the other two sides."

Converse (true) "If the square of the longest side of a triangle is equal to the sums of the squares of the other two sides, then the triangle is a right triangle."

Statement (true) "If you're in medical school now, then you had high grades in college."

Converse (false) "If you had high grades in college, then you're in medical school now."

CONVEX A set of points is convex if, for any two points in the set, all the points on the line segment joining them are also in the set. (See figure 30.) For contrast, see **concave**.

COORDINATES The coordinates of a point are a set of numbers that identify the location of that point. For example:

($x = 1, y = 2$) are Cartesian coordinates for a point in two-dimensional space.

($r = 3, \theta = 45°$) are polar coordinates for a point in two-dimensional space.

($x = 4, y = 5, z = 6$) are Cartesian coordinates for a point in three-dimensional space.

(latitude = 51 degrees north, longitude = 0 degree) are the terrestrial coordinates of the city of London.

(declination = −5 degree, 25 minutes, right ascension = 5 hours 33 minutes) are the celestial coordinates of the Great Nebula in Orion.

(See **Cartesian coordinates; polar coordinates.**)

COPLANAR A set of points is coplanar if they all lie in the same plane. Any three points are always coplanar. The vertices of a triangle are coplanar, but not the vertices of a pyramid. Two lines are coplanar if they lie in the same plane, that is, if they either intersect or are parallel.

COROLLARY A corollary is a statement that can be proved easily once a major theorem has been proved.

CORRELATION COEFFICIENT The correlation coefficient between two random variables X and Y (written as r or ρ) is defined to be:

$$r = \frac{Cov(X,Y)}{\sigma_X \sigma_Y}$$
$$= \frac{E(XY) - E(X)E(Y)}{\sqrt{[E(X^2) - (E(X))^2][E(Y^2) - (E(Y))^2]}}$$

$\text{Cov}(X, Y)$ is the covariance between X and Y;

σ_X and σ_Y are the standard deviations of X and Y, respectively; and E stands for expectation.

The correlation coefficient is always between -1 and 1. It tells whether or not there is a linear relationship between X and Y. If $Y = aX + b$, where a and b are constants and $a > 0$, then $r = 1$. If $a < 0$, then $r = -1$. If X and Y are almost, but not quite, linearly related, then r will be close to 1. If X and Y are completely independent, then $r = 0$.

Observations of two variables can be used to estimate the correlation between them. For some examples, see **regression**.

CORRESPONDING ANGLES (1) When a transversal cuts two lines, it forms four pairs of corresponding angles. In figure 31, angle 1 and angle 2 are a pair of corresponding angles. Angle 3 and angle 4 are another pair. In Euclidian geometry, if a transversal cuts two parallel lines, then the pairs of corresponding angles that are formed will be equal.

(2) When two polygons are congruent, or similar, each angle on one polygon is equal to a corresponding angle on the other polygon.

CORRESPONDING SIDES When two polygons are congruent, each side on one polygon is equal to a corresponding side on the other polygon. When two polygons are similar, the ratio of the length of a side on the big polygon to the length of its corresponding side on the little polygon is the same for all the sides.

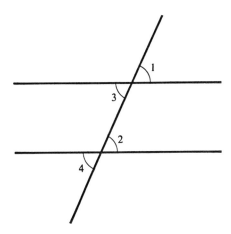

Figure 31 Corresponding angles

COSECANT The cosecant of θ is defined to be

$$\csc \theta = \frac{1}{\sin \theta}$$

(See **trigonometry**.)

COSINE The cosine of an angle θ in a right triangle is defined to be

$$\cos \theta = \frac{(adjacent\ side)}{(hypotenuse)}$$

The name comes from the fact that the cosine function is the cofunction for the sine function, because $\cos(\pi/2 - \theta) = \sin \theta$. The graph of the cosine function is periodic with an amplitude of 1 and a period of 2π. (See **trigonometry**.)

The table gives some special values of $\cos\theta$:

θ (degrees)	θ (radians)	$\cos\theta$
0	0	1
30	$\pi/6$	$\sqrt{3}/2$
45	$\pi/4$	$1/\sqrt{2}$
60	$\pi/3$	$1/2$
90	$\pi/2$	0
180	π	-1
270	$3\pi/2$	0
360	2π	1

The value of $\cos\theta$ can be found from the infinite series

$$\cos\theta = 1 - \frac{\theta^2}{2!} + \frac{\theta^4}{4!} - \frac{\theta^6}{6!} + \frac{\theta^8}{8!} - \cdots$$

COTANGENT The cotangent of θ (abbreviated ctn θ or cot θ) is defined to be

$$\text{ctn }\theta = \frac{1}{\tan\theta}$$

(See **trigonometry**.)

COTERMINAL Two angles are coterminal if they have the same terminal side when placed in standard position. (See **angle**.) For example, a 45° angle is coterminal with a 405° angle.

COUNTING NUMBERS The counting numbers are the same as the natural numbers: 1, 2, 3, 4, 5, 6, 7, ... They are the numbers you use to count something.

COVARIANCE The covariance of two random variables X and Y is a measure of how much X and Y

move together. The definition is

$$\text{Cov}(X, Y) = E[(X - E(x))(Y - E(Y))]$$

where E stands for expectation. If X and Y are completely independent, then $\text{Cov}(X, Y) = 0$. If Y is large at the same time that X is large, then $\text{Cov}(X, Y)$ will be large. However, if Y tends to be large when X is small, then the covariance will be negative. (See **correlation coefficient**.) The covariance can also be found from this expression:

$$\text{Cov}(X, Y) = E(XY) - E(X)E(Y)$$

CRAMER'S RULE Cramer's rule is a method for solving a set of simultaneous linear equations using determinants. For the 3×3 system:

$$a_1 x + b_1 y + c_1 z = k_1$$
$$a_2 x + b_2 y + c_2 z = k_2$$
$$a_3 x + b_3 y + c_3 z = k_3$$

The rule states:

$$x = \frac{\begin{vmatrix} k_1 & b_1 & c_1 \\ k_2 & b_2 & c_2 \\ k_3 & b_3 & c_3 \end{vmatrix}}{\begin{vmatrix} a_1 & b_1 & c_1 \\ a_2 & b_2 & c_2 \\ a_3 & b_3 & c_3 \end{vmatrix}}$$

$$y = \frac{\begin{vmatrix} a_1 & k_1 & c_1 \\ a_2 & k_2 & c_2 \\ a_3 & k_3 & c_3 \end{vmatrix}}{\begin{vmatrix} a_1 & b_1 & c_1 \\ a_2 & b_2 & c_2 \\ a_3 & b_3 & c_3 \end{vmatrix}}$$

$$z = \frac{\begin{vmatrix} a_1 & b_1 & k_1 \\ a_2 & b_2 & k_2 \\ a_3 & b_3 & k_3 \end{vmatrix}}{\begin{vmatrix} a_1 & b_1 & c_1 \\ a_2 & b_2 & c_2 \\ a_3 & b_3 & c_3 \end{vmatrix}}$$

The vertical lines symbolize determinant. (See **determinant**.)

To use Cramer's rule, first calculate the determinant of the whole matrix of coefficients. This determinant appears in the denominator of the solution for each variable. To calculate the numerator of the solution for x, set up the same matrix but make one substitution: cross out the column that contains the coefficients of x, and replace that column with the column of constants from the other side of the equal sign.

To use the rule to solve a system of n equations in n unknowns, you will have to calculate $n + 1$ determinants of dimension $n \times n$. This procedure could get tedious, but it is the kind of calculation that is well suited to be performed by a computer. For an example of the method, we can find the solution of this three-equation system:

$$\begin{aligned} 5x + y - 4z &= -1 \\ 3x - 6y + 2z &= -5 \\ 9x - y - 2z &= 13 \end{aligned}$$

The determinant in the denominator is

$$\begin{vmatrix} 5 & 1 & -4 \\ 3 & -6 & 2 \\ 9 & -1 & -2 \end{vmatrix} = -110$$

The three determinants in the numerators are

$$\begin{vmatrix} -1 & 1 & -4 \\ -5 & -6 & 2 \\ 13 & -1 & -2 \end{vmatrix} = -330$$

$$\begin{vmatrix} 5 & -1 & -4 \\ 3 & -5 & 2 \\ 9 & 13 & -2 \end{vmatrix} = -440$$

$$\begin{vmatrix} 5 & 1 & -1 \\ 3 & -6 & -5 \\ 9 & -1 & 13 \end{vmatrix} = -550$$

Then:

$$x = \frac{-330}{-110} = 3$$

$$y = \frac{-440}{-110} = 4$$

$$z = \frac{-550}{-110} = 5$$

CRITICAL POINT A critical point for a function is a point where the first derivative(s) is (are) zero. (See **extremum**.)

CRITICAL REGION If the calculated value of a test statistic falls within the critical region, then the null hypothesis is rejected. (See **hypothesis testing**.)

CROSS PRODUCT The cross product of two three-dimensional vectors

$$\mathbf{a} = (a_1, a_2, a_3) \quad \text{and} \quad \mathbf{b} = (b_1, b_2, b_3)$$

is:

$$\mathbf{a} \times \mathbf{b} = [(a_2 b_3 - a_3 b_2), (a_3 b_1 - a_1 b_3), (a_1 b_2 - a_2 b_1)]$$

$\mathbf{a} \times \mathbf{b}$ (read: $\mathbf{a}$ cross $\mathbf{b}$) is a vector with the following properties:

(1) $\|\mathbf{a} \times \mathbf{b}\| = \|\mathbf{a}\| \cdot \|\mathbf{b}\| \cdot \sin \theta_{ab}$

where θ_{ab} is the angle between $\mathbf{a}$ and $\mathbf{b}$ and $\|\mathbf{a}\|$ is the length of vector $\mathbf{a}$.

(2) $\mathbf{a} \times \mathbf{b}$ is perpendicular to both $\mathbf{a}$ and $\mathbf{b}$.

(3) The direction of $\mathbf{a} \times \mathbf{b}$ is determined by the right-hand rule: Put your right hand so that your fingers point in the direction from $\mathbf{a}$ to $\mathbf{b}$. Then your thumb points in the direction of $\mathbf{a} \times \mathbf{b}$. (See figure 32.)

(4) $\mathbf{a} \times \mathbf{b} = 0$ if $\mathbf{a}$ and $\mathbf{b}$ are parallel (i.e., if $\theta_{ab} = 0$).

(5) $\|\mathbf{a} \times \mathbf{b}\| = \|\mathbf{a}\| \cdot \|\mathbf{b}\|$ if $\mathbf{a}$ and $\mathbf{b}$ are perpendicular.

(6) The cross product is not commutative, since

$$\mathbf{a} \times \mathbf{b} = -\mathbf{b} \times \mathbf{a}$$

Here are some examples:

$$(2, 3, 4) \times (10, 15, 20) = (0, 0, 0)$$

Note the two vectors are parallel.

$$(4, 3, 0) \times (-3, 4, 0) = (0, 0, 25)$$

These two vectors are perpendicular, both with length 5, and they are both in the xy plane. Therefore, the cross product has length 25 and points in the direction of the z axis.

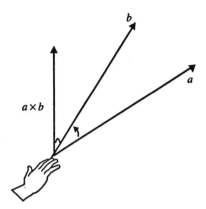

Figure 32   The right hand rule for the cross product

$$(-3, 4, 0) \times (4, 3, 0) = (0, 0, -25)$$

These are the same two vectors as in the previous example, except that the order of the cross product is reversed, so the resulting vector points in the opposite direction.

$$(0, 1, 0) \times (0, \sqrt{3}/2, 1/2) = (1/2, 0, 0)$$

These two vectors both have length 1, they are in the yz plane, and the angle between them is $30°$. Therefore, the cross product vector has length $1 \times 1 \times \sin 30° = 1/2$, and it points in the direction of the x axis.

The cross product is important in physics. The angular momentum vector $\mathbf{L}$ is defined by the cross product: $\mathbf{L} = \mathbf{r} \times \mathbf{p}$, where $\mathbf{p}$ is the linear momentum vector and $\mathbf{r}$ is the position vector.

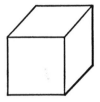

Figure 33   Cube

CUBE (1) A cube is a solid with six congruent square
faces. A cube can be thought of as a right prism
with square bases and four square lateral faces. (See
prism; polyhedron.) Dice are cubes and many ice
cubes are cubes. The volume of a cube with an edge
equal to a is a^3, which is read as "a cubed." The
surface area of a cube is $6a^2$. (See figure 33.)

(2) The cube of a number is that number raised
to the third power. For example, the cube of 2 is 8,
since $2^3 = 8$.

CUBE ROOT The cube root of a number is the number that, when multiplied together three times, gives
that number. For example, 4 is the cube root of 64,
since $4^3 = 4 \times 4 \times 4 = 64$. The cube root of x is
symbolized by $\sqrt[3]{x}$ or $x^{1/3}$.

CUBIC A cubic equation is a polynomial equation of
degree 3.

CUMULATIVE DISTRIBUTION FUNCTION
A cumulative distribution function gives the probability that a random variable will be less than or
equal to a specific value. (See **random variable.**)

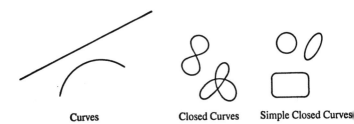

Curves   Closed Curves   Simple Closed Curves

Figure 34

CURL The curl of a three-dimensional vector field **f** (written as $\nabla \times \mathbf{f}$) is defined to be the vector

$$\nabla \times \mathbf{f} = \left(\left(\frac{\partial f_z}{\partial y} - \frac{\partial f_y}{\partial z} \right), \left(\frac{\partial f_x}{\partial z} - \frac{\partial f_z}{\partial x} \right), \left(\frac{\partial f_y}{\partial x} - \frac{\partial f_x}{\partial y} \right) \right)$$

It can be thought of as the cross product of the operator ∇ (del) with the field **f**. For application, see **Stokes' theorem; Maxwell's equations.**

CURVE A curve can be thought of as the path traced out by a point if it is allowed to move around space. A straight line is one example of a curve. A curve can have either infinite length, such as a parabola, or finite length, such as the ones shown in figure 34. If a curve completely encloses a region of a plane, it is called a closed curve. If a closed curve does not cross over itself, then it is a simple closed curve. A circle and an ellipse are both examples of simple closed curves.

CYCLOID If a wheel rolls along a flat surface, a point on the wheel traces out a multiarch curve known as a

Figure 35 Cycloid

cycloid. (See figure 35.) The cycloid can be defined
by the parametric equations

$$x = x_0 + a(\theta - \sin\theta), \; y = y_0 + a(1 - \cos\theta)$$

One important use of the cycloid is based on the
fact that, if a ball is to roll from uphill point A to
downhill point B, it will reach B the fastest along a
cycloid-shaped ramp. (See **calculus of variations**.)

CYLINDER A circular cylinder is formed by the
union of all line segments that connect corresponding points on two congruent circles lying in parallel planes. The two circular regions are the bases.
The segment connecting the centers of the two circles is called the axis. If the axis is perpendicular
to the planes containing the circles, then the cylinder is called a right circular cylinder. The distance
between the two planes is called the altitude of the
cylinder. The volume of a cylinder is the product of
the base area times the altitude. A soup can is one
example of a cylindrical object. (See figure 36.)

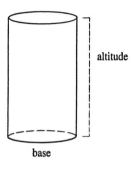

Figure 36 Cylinder

D

DECAGON A decagon is a polygon with 10 sides. A regular decagon has 10 equal sides and 10 angles, each of measure 144°. (See figure 37.)

DECIMAL NUMBERS The common way of representing numbers is by a decimal, or base-10, number system, wherein each digit represents a multiple of a power of 10. The position of a digit tells what power of 10 it is to be multiplied by. For example:

$$32,456 = 3 \times 10^4 + 2 \times 10^3 + 4 \times 10^2 + 5 \times 10^1 + 6 \times 10^0$$

We are so used to thinking of decimal numbers that we usually think of the decimal representation of the number as being the number itself. It is possible, though, to use other bases for number systems. Computers often use base-2 numbers (see **binary numbers**), and the ancient Babylonians used base-60 numbers.

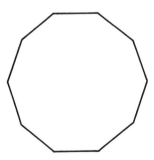

Figure 37 Decagon

A decimal fraction is a number in which the digits to the right of the decimal point are to be multiplied by 10 raised to a negative power:

$$32.564 = 3 \times 10^1 + 2 \times 10^0 + 5 \times 10^{-1} + 6 \times 10^{-2} + 4 \times 10^{-3}$$
$$= 32 + \frac{5}{10} + \frac{6}{100} + \frac{4}{1000} .$$

DECREASING FUNCTION A function $f(x)$ is a decreasing function if $f(a) < f(b)$ when $a > b$.

DEDUCTION A deduction is a conclusion arrived at by reasoning.

DEFINITE INTEGRAL If $f(x)$ represents a function of x that is always nonnegative, then the definite integral of $f(x)$ between a and b represents the area under the curve $y = f(x)$, above the x-axis, to the right of the line $x = a$, and to the left of the line $x = b$. (See figure 38.) The definite integral is represented by the expression

$$\int_a^b f(x)dx$$

where $\int$ is the integral sign, and a and b are the limits of integration.

The value of the definite integral can be found from the formula $F(b) - F(a)$, where F is an antiderivative function for f (that is, $dF/dx = f(x)$).

For contrast, see **indefinite integral**.

For example, we can find the area under one arch of the curve $y = \sin x$, from $x = 0$ to $x = \pi$.

$$area = \int_0^\pi \sin x dx$$

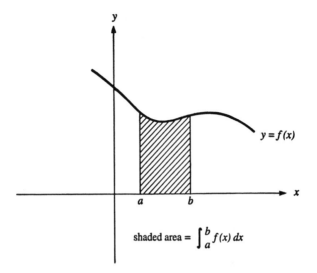

shaded area = $\int_a^b f(x)\,dx$

Figure 38 Definite Integral

The antiderivative function is $-\cos x$. Once the antiderivative has been found, it is customary to write the limits of integration next to a vertical line:

$$area = \cos x|_0^\pi = (-\cos \pi) - (-\cos 0)$$
$$= (-1) - (-1) = 2$$

Therefore, the total area under the curve is 2.

In cases where it is not possible to find an antiderivative function, see **numerical integration.**

If $f(x)$ is negative everywhere between a and b, then the value of the definite integral will be the negative of the area above the curve $y = f(x)$, below the x-axis, and between $x = a$ and $x = b$.

If $f(x)$ is positive in some places and negative in others, then the value of the definite integral will be the total area under the positive part of the curve minus the total area above the negative part of the curve.

Definite integrals can also be used to find other quantities. (See **arc length; volume, figure of revolution; surface area, figure of revolution; centroid.**)

DEGREE (1) A degree is a unit of measure for angles. One degree is equal to $1/360$ of a full rotation. The symbol for degree is a little raised circle, $°$. A full turn measures $360°$. A half turn measures $180°$. A quarter turn (a right angle) measures $90°$. (See **angle; radian measure.**)

(2) The degree of a polynomial is the highest power of the variable that appears in the polynomial. (See **polynomial.**)

DEL The del symbol ∇ is used to represent this vector of differential operators:

$$\nabla = (\frac{\partial}{\partial x}, \frac{\partial}{\partial y}, \frac{\partial}{\partial z})$$

(See **gradient; divergence; curl.**)

DELTA The Greek capital letter delta, which has the shape of a triangle: Δ, is used to represent "change in." For example, the expression Δx represents "the change in x." (See **calculus.**)

DE MOIVRE'S THEOREM De Moivre's theorem tells how to find the exponential of an imaginary number:

$$e^{i\theta} = \cos\theta + i\sin\theta$$

Note that θ is measured in radians. For example:

$$
\begin{aligned}
e^0 &= \cos 0 + i \sin 0 = 1 \\
e^{i\pi/2} &= \cos(\pi/2) + i \sin(\pi/2) = i \\
e^{i\pi} &= \cos \pi + i \sin \pi = -1
\end{aligned}
$$

To see why the theorem is reasonable, consider $(e^{ix})^2$. This expression should equal e^{2ix}, according to the laws of exponents. We can assume that the theorem is true and show that it is consistent with the law of exponents:

$$
\begin{aligned}
(e^{ix})^2 &= (\cos x + i \sin x)^2 \\
&= \cos^2 x + 2i \sin x \cos x - \sin^2 x \\
&= \cos 2x + i \sin 2x \\
&= e^{2ix}
\end{aligned}
$$

The theorem can also be shown by looking at the series expansion of e^{ix}:

$$
\begin{aligned}
e^{ix} &= 1 + ix - \frac{x^2}{2!} - \frac{ix^3}{3!} + \frac{x^4}{4!} + \cdots \\
&= \left(1 - \frac{x^2}{2!} + \frac{x^4}{4!} - \frac{x^6}{6!} + \cdots \right) \\
&\quad + i \left(x - \frac{x^3}{3!} + \frac{x^5}{5!} - \frac{x^7}{7!} + \cdots \right)
\end{aligned}
$$

The two series in parentheses are the series expansions for $\cos x$ and $\sin x$, so

$$
e^{ix} = \cos x + i \sin x
$$

This theorem plays an important part in the solution of some differential equations.

DE MORGAN Augustus De Morgan (1806 to 1871) was an English mathematician who studied logic. (See **De Morgan's laws**.)

DE MORGAN'S LAWS De Morgan's laws determine how the connectives AND, OR, and NOT interact in symbolic logic:

(NOT p)AND(NOTq) is equivalent to NOT(pORq)

(NOT p)OR(NOTq) is equivalent to NOT(pANDq)

In these expressions, p and q represent any sentences that have truth values (in other words, are either true or false). For example, the sentence "She is not rich and famous" is the same as the sentence "She is not rich, or else she is not famous."

DENOMINATOR The denominator is the bottom part of a fraction. In the fraction $\frac{2}{3}$, 3 is the denominator and 2 is the *numerator*. (To keep the terms straight, you might remember that "denominator" starts with "d," the same as "down.") If a fraction measures an amount of pie, the denominator tells how many equal slices the pie has been cut into. (See figure 39.) The numerator tells you how many slices you have.

DENSITY FUNCTION See **random variable**.

DEPENDENT VARIABLE The dependent variable stands for the output number of a function. In the equation $y = f(x)$, y is the dependent variable and x is the independent variable. The value of y depends on the value of x. You are free to choose any value of x that you wish (so long as it is in the domain of the function), but once you have chosen

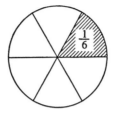

Figure 39

x the value of y is determined by the function. (See **function**.)

DERIVATIVE The derivative of a function is the rate of change of that function. On the graph of the curve $y = f(x)$, the derivative at x is equal to the slope of the tangent line at the point $(x, f(x))$. (See figure 40.)

If the function represents the position of an object as a function of time, then the derivative represents the velocity of the object. Derivatives can be calculated from this expression:

function: $y = f(x)$,
derivative:

$$y' = f'(x) = \frac{dy}{dx} = \lim_{\Delta x \to 0} \frac{f(x + \Delta x) - f(x)}{\Delta x}$$

Several rules are available that tell how to find the derivatives of different functions (c and n are constants):

$y = c$ $y' = 0$
$y = cx$ $y' = c$

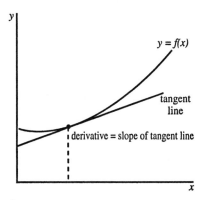

Figure 40

Sum Rule
$$y = f(x) + g(x) \qquad y' = f'(x) + g'(x)$$

Product Rule
$$y = f(x) \times g(x) \qquad y' = f(x)g'(x) + f'(x)g(x)$$

Power Rule
$$y = cx^n \qquad y' = cnx^{n-1}$$

Chain Rule
$$y = g(f(x)) \qquad y' = \frac{dg}{df}\frac{df}{dx} = \frac{dg}{dx}$$

Trigonometry
$$y = \sin x \qquad y' = \cos x$$
$$y = \cos x \qquad y' = -\sin x$$
$$y = \tan x \qquad y' = \sec^2 x$$
$$y = \operatorname{ctn} x \qquad y' = -\csc^2 x$$
$$y = \sec x \qquad y' = \sec x \tan x$$

$$y = \csc x \qquad y' = -\csc x \ \text{ctn} \ x$$
$$y = \arcsin x \qquad y' = (1 - x^2)^{-1/2}$$
$$y = \arctan x \qquad y' = (1 + x^2)^{-1}$$

Exponential
$$y = a^x \qquad y' = (\ln a)a^x$$

Natural Logarithm
$$y = \ln x \qquad y' = 1/x$$

(See also **implicit differentiation**.)

If y is a function of more than one independent variable, see **partial derivative**.

The derivative of the derivative is called the second derivative, written as $y''(x)$ or d^2y/dx^2. When the first derivative is positive, the curve is sloping upward. When the second derivative is positive, the curve is oriented so that it is concave upward. (See **extremum**.)

DESCARTES Rene Descartes (1596 to 1650) was a French mathematician and philosopher who is noted for the sentence "I think, therefore I am" and for developing the concept now known as rectangular, or **Cartesian coordinates**.

DESCRIPTIVE STATISTICS Descriptive statistics is the study of ways to summarize data. For example, the mean, median, and standard deviation are descriptive statistics that summarize some of the properties of a list of numbers. For contrast, see **statistical inference**.

DETERMINANT The determinant of a matrix is a number that is useful in describing the characteristics of the matrix. The determinant is symbolized by enclosing the matrix in vertical lines. The determinant of a 2×2 matrix is

$$\begin{vmatrix} a & b \\ c & d \end{vmatrix} = ad - bc$$

The determinant of a 3×3 matrix can be found from:

$$\begin{vmatrix} a_1 & b_1 & c_1 \\ a_2 & b_2 & c_2 \\ a_3 & b_3 & c_3 \end{vmatrix}$$

$$= a_1 \begin{vmatrix} b_2 & c_2 \\ b_3 & c_3 \end{vmatrix} - b_1 \begin{vmatrix} a_2 & c_2 \\ a_3 & c_3 \end{vmatrix} + c_1 \begin{vmatrix} a_2 & b_2 \\ a_3 & b_3 \end{vmatrix}$$

$$= a_1 b_2 c_3 + b_1 c_2 a_3 + c_1 a_2 b_3 - c_1 b_2 a_3 - a_1 c_2 b_3 - b_1 a_2 c_3$$

The 3×3 determinant consists of three terms. Each term contains an element of the top row multiplied by its minor. The minor of an element of a matrix can be found in this way: First, cross out all the elements in its row. Then cross out all the elements in its column. Then take the determinant of the 2×2 matrix consisting of all the elements that are left.

Note that the signs alternate, starting with plus for the element in the upper left hand corner.

For example:

$$\begin{vmatrix} 2 & 7 & 4 \\ 9 & 6 & 8 \\ 5 & 1 & 3 \end{vmatrix} = 2 \begin{vmatrix} 6 & 8 \\ 1 & 3 \end{vmatrix} - 7 \begin{vmatrix} 9 & 8 \\ 5 & 3 \end{vmatrix} + 4 \begin{vmatrix} 9 & 6 \\ 5 & 1 \end{vmatrix}$$

$$= 2(6 \cdot 3 - 8 \cdot 1) - 7(9 \cdot 3 - 5 \cdot 8) + 4(9 \cdot 1 - 5 \cdot 6)$$

$$= 2 \cdot 10 - 7 \cdot (-13) + 4 \cdot (-21) = 27$$

To find the determinant, you don't have to expand along the first row. Expansion along any row or column will produce the same value. If there is any row or column that contains many zeros, it is usually easiest to expand along that row (or column). For example:

$$\begin{vmatrix} 1 & 1 & 0 \\ 4 & 6 & 0 \\ 2 & 5 & 3 \end{vmatrix} = 0 \begin{vmatrix} 4 & 6 \\ 2 & 5 \end{vmatrix} - 0 \begin{vmatrix} 1 & 1 \\ 2 & 5 \end{vmatrix} + 3 \begin{vmatrix} 1 & 1 \\ 4 & 6 \end{vmatrix}$$

$$= 3(6 - 4) = 6$$

In this case we expanded along the last column.

There is no simple formula for determinants larger than 3×3, but the same method of expansion along a column or row may be used. One useful fact is that the value of the determinant will remain unchanged if you add a multiple of one row (or column) to another row (or column). By careful use of this trick, you can usually create a row consisting mostly of zeros, thus making it easier to evaluate the determinant. Even so, evaluation of large determinants is best left to a computer.

If the determinant is zero, then the matrix cannot be inverted. (See **inverse matrix.**) Some other properties of determinants are as follows:

$$\det(\mathbf{AB}) = \det \mathbf{A} \det \mathbf{B}$$

$\det \mathbf{I} = 1$ (**I** is the identity matrix),

$\det \mathbf{A}^{-1} = 1/\det \mathbf{A}$

Determinants can be used to solve simultaneous linear equation systems. (See **Cramer's rule.**)

DIAGONAL A diagonal is a line segment connecting two nonadjacent vertices of a polygon. For example, a rectangle has two diagonals, each connecting a pair of opposite corners.

DIAMETER The diameter of a circle is the length of a line segment joining two points on the circle and passing through the center. The term diameter can also mean the segment itself. The diameter is equal to twice the radius, and $d = c/\pi$, where c is the circumference. The diameter is the longest possible distance across the circle. Our Milky Way galaxy is shaped like a disk with a bulge in the middle. The diameter of the circle that makes up the outer edge of the disk is about 100,000 light-years.

The diameter of a sphere is the length of a line segment joining two points on the sphere and passing through the center. The sun is a sphere with a diameter of about 865,000 miles.

DIFFERENCE The difference between two numbers is the result obtained by subtracting them. In the equation $5 - 3 = 2$, the number 2 is the difference. If two points are located along a number line, then the absolute value of their difference will be the distance between them. For example, Bridgeport is at mile 28 of the Connecticut Turnpike, and Stamford is at mile 7. The distance between them is the difference: $28 - 7 = 21$ miles.

DIFFERENCE OF TWO SQUARES An expression is a difference of two squares if it is of the form $a^2 - b^2$. This expression can be factored as follows:

$$a^2 - b^2 = (a - b)(a + b)$$

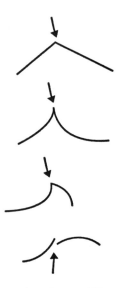

Figure 41 Curves that are not differentiable at the point marked by the arrow

DIFFERENTIABLE A continuous function is differentiable over an interval if its derivative exists everywhere in that interval. (See **calculus; derivative.**) This means that the graph of the function is smooth, with no kinks, cusps, or breaks. (See figure 41.)

DIFFERENTIAL Differential refers to an infinitesimal change in a variable. It is symbolized by d, as in dx. The derivative dy/dx can be thought of as a ratio of two differential changes. (See **derivative.**)

DIFFERENTIAL EQUATION A differential equation is an equation containing the derivatives of a function with respect to one or more independent variables. The *order* of the equation is the highest derivative that appears; for example, the equation

$$\frac{dy}{dx} - f(x) = 0$$

is a first-order equation, which can be solved by turning it into an integral:

$$(1) \quad y = \int f(x)dx$$

Second-order equations appear commonly in physics, since force equals mass times acceleration. If you know an equation for the force acting on a particle that moves in one dimension, its position x at a time t will be found by solving this differential equation:

$$(2) \quad F = m\frac{d^2x}{dt^2}$$

Note that, in the above equation, t is now the independent variable, and x is the dependent variable. For example, the motion of a weight connected to a spring is given by this equation:

$$(3) \quad m\frac{d^2x}{dt^2} = -kx$$

where m is the mass and k is a constant depending on the nature of the spring. The solution is:

$$x = A\sin(\omega t + B)$$

where ω is defined to be $\sqrt{k/m}$, and A and B are two arbitrary constants whose value depends on the initial position and velocity of the weight. Note that solving an integral, or first-order differential equation, results in one arbitrary constant. When solving a second-order differential equation, there will be two arbitrary constants in the solution.

Equation (3) can be generalized to the form:

$$(4) \quad \frac{d^2x}{dt^2} + c_1 \frac{dx}{dt} + c_0 x = 0$$

where the term involving dx/dt represents the friction acting on the weight. A similar type of equation describes the behavior of oscillating electric circuits. The solution is given by:

$$x = B_1 e^{r_1 t} + B_2 e^{r_2 t}$$

where B_1 and B_2 are the two arbitrary constants, and r_1 and r_2 are the solutions of the quadratic equation

$$r^2 + c_1 r + c_0 = 0$$

If the two values for r are pure imaginary numbers, then the solution will oscillate. This comes from **De Moivre's theorem:**

$$e^{i\theta} = \cos\theta + i\sin\theta$$

If the two values for r are real, then the result will be an exponential function. If the two values for r are complex numbers (call them $r_0 + i\omega$ and $r_0 - i\omega$), then the solution will be a mixture of oscillating and exponential factors as follows:

$$x = e^{r_0 t}(B_1 \sin\omega t + B_2 \cos\omega t)$$

where again B_1 and B_2 are the arbitrary constants.

Equation (4) above is called a *linear* differential equation. The general form of a second-order linear differential equation is:

$$(5) \quad \left[\frac{d^2}{dt^2} + f_1(t)\frac{d}{dt} + f_0(t)\right] x = f(t)$$

The equation is said to be *homogeneous* if the right hand side function is zero; in other words, it can be written in the form:

$$(6) \quad \left[\frac{d^2}{dt^2} + f_1(t)\frac{d}{dt} + f_0(t) \right] x = 0$$

If y_1 is a solution of equation (5), and y_0 is a solution of equation (6), then $y_1 + y_0$ will also be a solution of (5).

All of the above equations are called *ordinary* differential equations because there is only one independent variable. If the equation contains derivatives with respect to more than one independent variable, then it is called a *partial* differential equation. For example, the equation

$$\frac{\partial^2 f}{\partial x^2} = \frac{1}{v^2} \frac{\partial^2 f}{\partial t^2}$$

is a second-order partial differential equation. An example of a solution is $f(x,t) = \sin(x - vt)$, which defines a wave moving in one spatial dimension x, where t is time and v is the velocity of the wave.

DIFFERENTIATION Differentiation is the process of finding a derivative. (See **derivative**.)

DIGIT The digits are the 10 symbols 0, 1, 2, 3, 4, 5, 6, 7, 8, 9. For example, 1462 is a four-digit number, and the number 3.46 contains two digits to the right of the decimal point. There are 10 digits in the commonly used decimal system. In the binary system only two digits are used. (See **binary numbers**.)

DIGITAL A digital system is a system where numerical quantities are represented by a device that shifts between discrete states, rather than varying continuously. For example, an abacus is an example of a digital device, because numbers are represented by beads that are either "up" or "down"; there is no meaning for a bead that is partway up or down. Pocket calculators and modern computers are also digital devices. A digital device can be more accurate than an analog device because the system only needs to distinguish between clearly separated states; it is not necessary to make fine measurements. Other examples of digital devices include clocks that display numbers to represent the time (rather than show hands moving around a circle) and music stored as a series of numbers on compact discs. For contrast, see **analog**.

DIHEDRAL ANGLE A dihedral angle is the figure formed by two intersecting planes. Consider two intersecting lines, one in each plane, that are both perpendicular to the line formed by the intersecting planes. Then the angle between these two lines is the size of the dihedral angle. (See figure 42.)

DIMENSION The dimension of a space is the number of coordinates needed to identify a location in that space. For example, a line is one dimensional; a plane is two dimensional; and the space we live in is three dimensional.

DIRECTION COSINES The direction cosines of a line are the cosines of the angles that the line makes with the three coordinate axes.

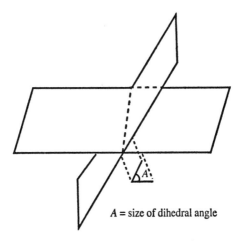

A = size of dihedral angle

Figure 42   Dihedral angle

DIRECTLY PROPORTIONAL If y and x are related by an equation of the form $y = kx$, where k is a constant, then y is said to be directly proportional to x.

DIRECTRIX A directrix is a line that helps to define a geometric figure. In particular, a directrix line is used in the definition of conic sections.

DISCRETE Discrete refers to a situation where the possibilities are distinct and separated from each other. For example, the number of people in a city is discrete, because there is no such thing as a fractional person. Measurements of time and distance, however, are not discrete, because they can vary over a continuous range. (See **continuous**.) Measurements of the energy levels of electrons in quantum

mechanics are discrete, because there are only a few
possible values for the energy.

DISCRETE RANDOM VARIABLE A discrete
random variable is a random variable which can only
take on values from a discrete list. The probability
function (or density function) lists the probability
that the variable will take on each of the possible
values. The sum of the probabilities for all of the
possible values must be 1.

For examples, see **binomial distribution; Poisson distribution; geometric distribution; hypergeometric distribution.** For contrast, see
continuous random variable.

DISCRIMINANT The discriminant (D) of a quadratic equation $ax^2+bx+c=0$ is $D=b^2-4ac$. If a, b,
and c are real numbers, the discriminant allows you
to determine the characteristics of the solution for
x. If D is a positive perfect square, then x will have
two rational values. If $D=0$, then x will have one
rational solution. If D is positive but is not a perfect
square, then x will have two irrational solutions. If D
is negative, then x will have two complex solutions.
(See **quadratic equation.**)

DISJOINT Two sets are disjoint if they have no elements in common, that is, if their intersection is
the empty set. The set of triangles and the set of
quadrilaterals are disjoint.

DISJUNCTION A disjunction is an OR statement
of the form: "A OR B." It is true if either A or B
is true.

DISTANCE The distance postulate states that for every two points in space there exists a unique positive number that can be called the distance between these two points. The distance between point A and point B is often written as AB. If $A = (a_1, a_2, a_3)$ and $B = (b_1, b_2, b_3)$, then the distance between them can be found from the distance formula (which is based on the Pythagorean theorem):

$$AB = \sqrt{(a_1 - b_1)^2 + (a_2 - b_2)^2 + (a_3 - b_3)^2}$$

DISTRIBUTIVE PROPERTY The distributive property says that $a(b + c) = ab + ac$ for all a, b, and c. For example,

$$
\begin{aligned}
3(4 + 5) &= 3 \cdot 4 + 3 \cdot 5 \\
3 \cdot 9 &= 12 + 15 \\
27 &= 27.
\end{aligned}
$$

DIVERGENCE The divergence of a vector field $\mathbf{f}$ (written as $\nabla \cdot \mathbf{f}$) is defined to be the scalar

$$\nabla \cdot \mathbf{f} = \frac{\partial f_x}{\partial x} + \frac{\partial f_y}{\partial y} + \frac{\partial f_z}{\partial z}$$

It can be thought of as the dot product of the operator ∇ (del) with the field $\mathbf{f}$. For application, see **Maxwell's equations.**

DIVERGENCE THEOREM The divergence theorem states that if $\mathbf{E}$ is a three-dimensional vector field, then the surface integral of $\mathbf{E}$ over a closed surface is equal to the triple integral of the divergence of $\mathbf{E}$ over the volume enclosed by that surface:

$$\iint_{surface\ S} \mathbf{E}\cdot d\mathbf{S} = \iiint_{interior\ of\ S} (\nabla\cdot\mathbf{E})\, dV$$

For application, see **Maxwell's equations**.

DIVERGENT SERIES A divergent series is an infinite series with no finite sum. A series that does have a finite sum is called a **convergent series**.

DIVIDEND In the equation $a \div b = c$, a is called the dividend.

DIVISION Division is the opposite operation of multiplication. If $a \times b = c$, then $c \div b = a$. For example, $6 \times 8 = 48$, and $48 \div 6 = 8$. The symbol "$\div$" is used to represent division in arithmetic. In algebra most divisions are written as fractions: $b \div a = b/a$. For computational purposes, b/a is symbolized by $a\overline{)b}$.

DIVISOR In the equation $a \div b = c$, b is called the divisor.

DODECAHEDRON A dodecahedron is a polyhedron with 12 faces. (See **polyhedron.**) (See figure 43.)

DOMAIN The domain of a function is the set of all possible values for the argument (the input number) of the function. (See **function.**)

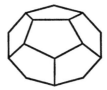

Figure 43   Dodecahedron

DOT PRODUCT Let **a** and **b** be two n-dimensional vectors, whose components are:

$$\begin{aligned} \mathbf{a} &= (a_1, a_2, a_3, \ldots a_n) \\ \mathbf{b} &= (b_1, b_2, b_3, \ldots b_n) \end{aligned}$$

The dot product of the two vectors (written as $\mathbf{a} \cdot \mathbf{b}$) is defined to be:

$$\mathbf{a} \cdot \mathbf{b} = a_1 b_1 + a_2 b_2 + a_3 b_3 + \cdots + a_n b_n$$

To find a dot product, you multiply all the corresponding components of each vector and then add together all of these products. In two-dimensional space this becomes:

$$\begin{aligned} \mathbf{a} &= (a_x, a_y) \\ \mathbf{b} &= (b_x, b_y) \\ \mathbf{a} \cdot \mathbf{b} &= a_x b_x + a_y b_y \end{aligned}$$

The dot product is a number, or scalar, rather than a vector. The dot product is also called the scalar product. Another form for the dot product can be found by defining the length of each vector:

$$r_a = \sqrt{a_x^2 + a_y^2}, \ r_b = \sqrt{b_x^2 + b_y^2}$$

Then:

$$\mathbf{a} \cdot \mathbf{b} = r_a r_b \left(\frac{a_x b_x}{r_a r_b} + \frac{a_y b_y}{r_a r_b} \right)$$

Let θ_a be the angle between vector **a** and the x-axis, θ_b be the angle between vector **b** and the x-axis, and $\theta = \theta_a - \theta_b$ be the angle between the two vectors. Then:

$$\frac{a_x}{r_a} = \cos \theta_a; \ \frac{b_x}{r_b} = \cos \theta_b$$

$$\frac{a_y}{r_a} = \sin\theta_a; \quad \frac{b_y}{r_b} = \sin\theta_b$$

We can rewrite the dot product formula:

$$\mathbf{a} \cdot \mathbf{b} = r_a r_b (\cos\theta_a \cos\theta_b + \sin\theta_a \sin\theta_b)$$

Using the formula for the cosine of the difference between two angles gives

$$\mathbf{a} \cdot \mathbf{b} = r_a r_b \cos\theta$$

The last formula says that the dot product can be found by multiplying the magnitude of the two vectors and the cosine of the angle between them. This means that the dot product is already good for two things:

1. Two nonzero vectors will be perpendicular if and only if their dot product is zero. (A zero dot product means that $\cos\theta = 0$, meaning $\theta = 90°$.)

2. The dot product $\mathbf{a} \cdot \mathbf{b}$ can be used to find the projection of vector $\mathbf{a}$ on vector $\mathbf{b}$:

$$\text{projection of } \mathbf{a} \text{ on } \mathbf{b} = \mathbf{P} = \frac{\mathbf{a} \cdot \mathbf{b}}{\mathbf{b} \cdot \mathbf{b}} \mathbf{b}$$

Note that the quantity $(\mathbf{a}\cdot\mathbf{b})/(\mathbf{b}\cdot\mathbf{b})$ is a scalar, so the projection vector is formed by multiplying a scalar times the vector $\mathbf{b}$. (See figure 44.)

Here is an example of how the dot product can be used to find the angle between two vectors. The cosine of the angle between the vectors $(1,1)$ and $(2,4)$ will be given by

$$\cos\theta = \frac{1\cdot 2 + 1\cdot 4}{\sqrt{2}\cdot\sqrt{20}} = 0.95$$

$$\theta = \arccos.95 = 18°$$

DOUBLE INTEGRAL The double integral of a two-variable function $f(x,y)$ represents the volume

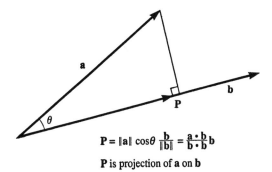

$$\mathbf{P} = \|\mathbf{a}\| \cos\theta \, \frac{\mathbf{b}}{\|\mathbf{b}\|} = \frac{\mathbf{a} \cdot \mathbf{b}}{\mathbf{b} \cdot \mathbf{b}} \mathbf{b}$$

P is projection of **a** on **b**

Figure 44

under the surface $z = f(x,y)$ and above the xy plane in a specified region. For example:

$$\int_{y=c}^{y=d} \int_{x=a}^{x=b} f(x,y) \, dx \, dy$$

represents the volume under the surface $z = f(x,y)$ over the rectangle from $x = a$ to $x = b$ and $y = c$ to $y = d$. (See figure 45.)

This assumes that $f(x,y)$ is nonnegative everywhere within the limits of integration. If $f(x,y)$ is negative, then the double integral will give the negative of the volume above the surface and below the x, y plane.

For example, consider a sphere of radius r with center at the origin. The equation of this sphere is

$$x^2 + y^2 + z^2 = r^2$$

The equation

$$z = f(x,y) = \sqrt{r^2 - x^2 - y^2}$$

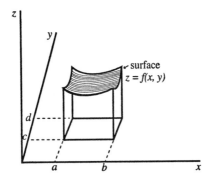

Figure 45 Double integral

defines a surface, which is the top half of the sphere. The volume below this surface and above the plane $z = 0$ is given by the double integral:

$$\int_{x=-r}^{x=r} \int_{y=-\sqrt{r^2-x^2}}^{y=\sqrt{r^2-x^2}} \sqrt{r^2 - x^2 - y^2} \; dy \, dx$$

The limits of integration for y will be from $-\sqrt{r^2 - x^2}$ to $\sqrt{r^2 - x^2}$, and the limits for x will be from $-r$ to r.

Evaluate the inner integral (involving y) first. While evaluating the inner integral, treat x as a constant. Define $A = \sqrt{r^2 - x^2}$, then use the trigonometric substitution $y = A \sin \theta$; $dy = A \cos \theta d\theta$; $\theta = \arcsin(y/A)$. Then the integral can be written:

$$\int_{y=-\sqrt{r^2-x^2}}^{y=\sqrt{r^2-x^2}} \sqrt{r^2 - x^2 - y^2} \, dy$$

$$= \int_{y=-A}^{y=A} \sqrt{A^2 - y^2} \, dy$$

$$= \int_{y=-A}^{y=A} \sqrt{A^2[1 - (y/A)^2]} \, dy$$

$$= \int_{\theta=\arcsin(-A/A)}^{\theta=\arcsin(A/A)} A\sqrt{1 - \sin^2 \theta} \, A \cos \theta \, d\theta$$

$$= A^2 \int_{\arcsin(-1)}^{\arcsin(1)} \cos^2 \theta \, d\theta$$

$$= A^2 \int_{-\pi/2}^{\pi/2} [\frac{1}{2}(1 + \cos 2\theta)] \, d\theta$$

$$= \frac{A^2}{2} \int_{-\pi/2}^{\pi/2} d\theta + \frac{A^2}{2} \int_{-\pi/2}^{\pi/2} \cos 2\theta \, d\theta$$

$$= \frac{A^2}{2} \theta|_{-\pi/2}^{\pi/2} - \frac{A^2}{4} \sin 2\theta|_{-\pi/2}^{\pi/2}$$

The second term in the integral is zero, so the result for the inner integral is:

$$\int_{y=-\sqrt{r^2-x^2}}^{y=\sqrt{r^2-x^2}} \sqrt{r^2 - x^2 - y^2}\ dy = \frac{\pi}{2}(r^2 - x^2)$$

Now substitute this expression in place of the inner integral, and then evaluate the outer integral involving x:

$$\int_{-r}^{r} \frac{\pi}{2}(r^2 - x^2)dx = \frac{2}{3}\pi r^3$$

(Note that this is half of the volume of a complete sphere.)

DYADIC OPERATION A dyadic operation is an operation that requires two operands. For example, addition is a dyadic operation. The logical operation AND is dyadic, but the logical operation NOT is not dyadic.

E

e The letter e is used to represent a fundamental irrational number with the decimal approximation $e = 2.7182818\ldots$. The letter e is the base of the natural logarithm function. (See **calculus; logarithm.**) The area under the curve $y = 1/x$ from $x = 1$ to $x = e$ is equal to 1.

The value of e can be found from this series:

$$e = 2 + \frac{1}{2!} + \frac{1}{3!} + \frac{1}{4!} + \frac{1}{5!} + \cdots$$

The value of e can also be found from the expression

$$e = \lim_{w \to 0}(1 + w)^{1/w}$$

In calculus, the function e^x is important because its derivative is itself: e^x.

ECCENTRICITY The eccentricity of a conic section is a number that indicates the shape of the conic section. The eccentricity (e) is the distance to the focal point divided by the distance to the directrix line. This ratio will be a constant, according to the definition. (See **conic section.**) If $e = 1$, then the conic section is a parabola; if $e > 1$, it is a hyperbola; and if $e < 1$, it is an ellipse.

The eccentricity of an ellipse measures how far the ellipse differs from being a circle. You can think of a circle as being normal (eccentricity $= 0$), with the ellipses becoming more and more eccentric as they become flatter. The eccentricity of the ellipse $x^2/a^2 + y^2/b^2 = 1$ is equal to

$$e = \frac{\sqrt{a^2 - b^2}}{a}$$

EDGE The edge of a polyhedron is a line segment
where two faces intersect. For example, a cube has
12 edges.

EIGENVALUE Suppose that a square matrix **A** multiplies a vector **x**, and the resulting vector is proportional to **x**:

$$\mathbf{A}\mathbf{x} = \lambda\mathbf{x}$$

In this case, λ is said to be an eigenvalue of the matrix **A**, and **x** is the corresponding eigenvector. In
order to find the eigenvalues, rewrite the equation
like this:

$$(\lambda\mathbf{I} - \mathbf{A})\mathbf{x} = \mathbf{0}$$

where **I** is the appropriate identity matrix. If the
inverse of the matrix $(\lambda\,\mathbf{I} - \mathbf{A})$ exists, then the
only solution is the trivial case $\mathbf{x} = \mathbf{0}$. However, if
the determinant is zero, then we will be able to find
nonzero vectors that meet our condition.

For example, let **A** be the matrix

$$\begin{pmatrix} 2 & 3 \\ 6 & 5 \end{pmatrix}$$

Set the determinant of $(\lambda\,\mathbf{I} - \mathbf{A})$ equal to zero:

$$\begin{vmatrix} \lambda - 2 & -3 \\ -6 & \lambda - 5 \end{vmatrix} = 0$$
$$(\lambda - 2)(\lambda - 5) - 18 = 0$$
$$\lambda^2 - 7\lambda - 8 = 0$$

From the quadratic formula, we find two values
for λ: 8 and -1. These are the two eigenvalues.

Now, to solve for the first eigenvector, set up this
matrix equation:

$$\begin{pmatrix} 2 & 3 \\ 6 & 5 \end{pmatrix}\begin{pmatrix} x \\ y \end{pmatrix} = \begin{pmatrix} 8x \\ 8y \end{pmatrix}$$

which is equivalent to this two-equation system:

$$2x + 3y = 8x$$

$$6x + 5y = 8y$$

These equations are equivalent, so there are an infinite number of solutions. This means that once one eigenvector has been found, any vector that is a multiple of that vector will also be an eigenvector. In this case let $x=1$, then we find $y = 2$. Therefore, any vector of the form $(x, 2x)$ is an eigenvector associated with the eigenvalue 8.

Using a similar procedure, we can find that the eigenvectors associated with the eigenvalue -1 are of the form $(x, -x)$.

When solving for the eigenvalues of an $n \times n$ matrix, you will have to solve a polynomial equation of degree n. This means there can be as many as n distinct solutions. Often the solutions will be complex numbers. As you can see, solving for eigenvalues of large matrices is a difficult problem.

Eigenvalues have many applications in fields such as quantum mechanics.

EIGENVECTOR See **eigenvalue.**

ELEMENT An element of a set is a member of the set.

ELLIPSE An ellipse is the set of all points in a plane such that the sum of the distances to two fixed points is a constant. Ellipses look like flattened circles. (See figure 46.) Each of these two fixed points is known as a *focus* or *focal point.* (The plural of focus is *foci.*) The longest distance across the ellipse is known as

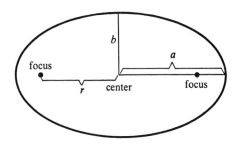

Figure 46 Ellipse

the major axis. (Half of this distance is known as the semimajor axis.) The shortest distance across is the minor axis.

The center of the ellipse is the midpoint of the segment that joins the two foci. The equation of an ellipse with center at the origin is

$$\frac{x^2}{a^2} + \frac{y^2}{b^2} = 1$$

where a is the length of the semimajor axis, and b is the length of the semiminor axis. The equation of an ellipse with center at point (h, k) is

$$\frac{(x-h)^2}{a^2} + \frac{(y-k)^2}{b^2} = 1$$

(This assumes the major axis is parallel to the x axis. To learn how to find the equation of an ellipse with a different orientation, see **rotation**.)

The area of an ellipse is $A = \pi ab$.

The shape of an ellipse can be characterized by a number that measures the degree of flattening,

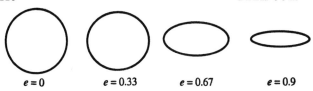

Figure 47 Ellipses

known as the eccentricity. The eccentricity (e) is

$$e = \frac{\sqrt{a^2 - b^2}}{a} = \frac{r}{a}$$

where r is the distance from the center to one of the focal points, as shown in figure 46. When $e = 0$, there is no flattening and the ellipse is the same as a circle. As e becomes larger and approaches 1, the ellipse becomes flatter and flatter. (See figure 47.)

An ellipse can also be defined as the set of points such that the distance to a fixed point divided by the distance to a fixed line is a constant that is less than 1. The constant is the eccentricity of the ellipse. (See **conic sections**.)

One reason why ellipses are important is that the path of an orbiting planet is an ellipse, with the sun at one focus. The orbit of the earth is an ellipse that is almost a perfect circle. Its eccentricity is only 0.017.

ELLIPSOID An ellipsoid is a solid of revolution formed by rotating an ellipse about one of its axes. If the ellipse has semimajor axis a and semiminor axis b, then the ellipsoid formed by rotating the ellipse about its major axis will have the volume $\frac{4}{3}\pi ab^2$.

EMPTY SET An empty set is a set that contains no elements. It is symbolized by ∅. For example, the set of all people over 100 feet tall is an example of an empty set.

EQUATION An equation is a statement that says that two mathematical expressions have the same value. The symbol $=$ means "equals," as in $4 \times 5 = 20$. If all the items in an equation are numbers, then the equation is an arithmetic equation and it is either true or false. For example, $10 + 15 = 35$ is true, but $2 + 2 = 5$ is false. If the equation contains a letter that represents an unknown number, then there will usually be some values of the unknown that make the equation true. For example, the equation $5 + 3 = x$ is true if x has the value 8; otherwise it is false. An equation in one unknown is said to be solved when it is written in the form $x = (expression)$, where *(expression)* depends only on numbers or on letters that stand for known quantities.

When solving an equation, the basic rule is: Whatever you do to one side of the equation, make sure you do exactly the same thing to the other side. For example, the equation $10x + 5 = 6x + 2$ can be solved by subtracting $6x + 2$ from both sides:

$$10x + 5 - (6x + 2) = 0$$
$$4x + 3 = 0$$

Subtract 3 from both sides:

$$4x = -3$$

Divide both sides by 4:

$$x = -\frac{3}{4}$$

You cannot divide both sides of an equation by zero, since division by zero is meaningless. It also does no good to multiply both sides by zero. Squaring both sides of an equation, or multiplying both sides by an expression that might equal zero, can sometimes introduce an extraneous root: a root that is a solution of the new equation but is not a solution of the original equation. For example, you might solve the equation $\sqrt{x^2 - 2x + 1} = 2x - 5$ by squaring both sides:

$$\begin{aligned} x^2 - 2x + 1 &= 4x^2 - 20x + 25 \\ 3x^2 - 18x + 24 &= 0 \\ x^2 - 6x + 8 &= 0 \\ (x-4)(x-2) &= 0 \end{aligned}$$

In this case $x = 4$ does satisfy the original equation, but $x = 2$ does not. This means that $x = 2$ is an extraneous root.

An equation that can be put in the general form $ax + b = 0$, where x is unknown and a and b are known, is called a **linear equation**. Any one-unknown equation can be written in this form provided that it contains no terms with x^2, $1/x$, or any term with x raised any power other than 1. An equation involving x^2 and x is called a **quadratic equation**, and can be written in the form $ax^2 + bx + c = 0$. For equations involving higher powers of x, see **polynomial**.

When an equation contains two unknowns, there will in general be many possible pairs of the unknowns that make the equation true. For example, $2x + y = 20$ will be satisfied by $(x = 0, y = 20)$; $(x = 5, y = 10)$; $(x = 10, y = 0)$; and many other pairs of values. In a case like this, you can often solve for one unknown as a function of the other, and you can draw a picture of the relationship between the unknowns. Also, you can find a unique solution for the two unknowns if you have two equations that must be true simultaneously. (See **simultaneous equations**.)

Another kind of equation is an equation that is true for all values of the unknown. This type of equation is called an identity. For example, $y^3 = y \times y \times y$ is true for every possible value of y. Usually it is possible to tell from the context the difference between a regular (or conditional) equation and an identity, but sometimes a symbol with three lines ($\equiv$) is used to indicate an identity: $\sin^2 x + \cos^2 x \equiv 1$.

The above equation is true for every possible value of x.

EQUILATERAL TRIANGLE An equilateral triangle is a triangle with three equal sides. All three of the angles in an equilateral triangle are 60° angles. The area of an equilateral triangle of side s is $s^2\sqrt{3}/4$.

EQUIVALENT Two logic sentences are equivalent if they will always have the same truth value. For example, the sentence "$p \to q$" ("IF p THEN q") is equivalent to the sentence "(NOT q) $\to$ (NOT p)."

EQUIVALENT EQUATIONS Two equations are equivalent if their solutions are the same. For example, the equation $x + 3y = 10$ is equivalent to the equation $2x + 6y = 20$.

1	2	3	4	5	6	7	8	9	10
11	12	13	14	15	16	17	18	19	20
21	22	23	24	25	26	27	28	29	30
31	32	33	34	35	36	37	38	39	40
41	42	43	44	45	46	47	48	49	50
51	52	53	54	55	56	57	58	59	60
61	62	63	64	65	66	67	68	69	70
71	72	73	74	75	76	77	78	79	80
81	82	83	84	85	86	87	88	89	90
91	92	93	94	95	96	97	98	99	100

Figure 48   Eratosthenes Sieve

ERATOSTHENES Eratosthenes of Cyrene (276 to 194 BC) was a Greek mathematician and astronomer who is the first person known to have calculated the circumference of the Earth. (See **Eratosthenes sieve.**)

ERATOSTHENES SIEVE Eratosthenes' sieve is a means for determining all of the prime numbers less than a given number by filtering out all of the non-prime numbers. Figure 48 illustrates all of the prime numbers less than 100. First, cross out all multiples of two after two. Then, cross out all multiples of three after three, then all multiples of five after five, and continue the process for all of the prime numbers below $\sqrt{100} = 10$.

ESTIMATOR An estimator is a quantity, based on observations of a sample, whose value is taken as an indicator of the value of an unknown population parameter. For example, the sample average $\bar{x}$ is often used as an estimator of the unknown population mean μ. (See **statistical inference**.)

EUCLID Euclid (c 300 BC) was a Greek mathematician who lived in Alexandria and is noted for his treatise on geometry, *Elements*, which focused on developing a logical structure with proofs. Much of the work is of the nature of a textbook based on work by earlier writers, but the completeness of the work made it one of the most influential mathematical works of all time. The geometry of our everyday world is still known as Euclidian geometry.

EUCLID'S ALGORITHM Euclid's algorithm provides a way of determining the greatest common factor of two natural numbers a and b. Assume $a > b$. First, calculate the remainder to the division $a \div b$ (call it r_1). Then, calculate the remainder to the division $b \div r_1$ (call it r_2); then calculate the remainder in the division $r_1 \div r_2$. Keep repeating the process, where at each stage you divide the remainder in the previous step by the new remainder, until you find a remainder of 0. Then, the last nonzero remainder that you found is the greatest common divisor of a and b.

For example, we can find the greatest common factor of 1683 and 714.

First division:

$$1683 \div 714 = 2 \text{ remainder } 255$$

Second division:

$$714 \div 255 = 2 \text{ remainder } 204$$

Third division:

$$255 \div 204 = 1 \text{ remainder } 51$$

Fourth division:

$$204 \div 51 = 4 \text{ remainder } 0$$

Since 51 is the last nonzero remainder, it is the greatest common factor of 1683 and 714.

EUCLIDIAN GEOMETRY Euclidian geometry is the geometry based on the postulates of Euclid. Euclidian geometry in three-dimensional space corresponds to our intuitive ideas of what space is like. For contrast, see **non-Euclidian geometry**.

EULER Leonhard Euler (1707 to 1783), a Swiss mathematican who worked much of his life in St. Petersburg and Berlin, advanced mathematical ideas in many areas, including analytic geometry, calculus, trigonometry, the theory of complex numbers, and number theory. He also is responsible for much mathematical notation that is now common, including $\sum$ for summation, e for the base of the natural logarithms, $f()$ for functions, π for the circumference of a circle of diameter 1, and i for $\sqrt{-1}$.

EVEN FUNCTION The function $f(x)$ is an even function if it satisfies the property that $f(x) = f(-x)$. For example, $f(x) = \cos x$ and $g(x) = x^2$ are both even functions. For contrast, see **odd function**.

EVEN NUMBER An even number is a natural number that is divisible evenly by 2. For example, 2, 4, 6, 8, 10, 12, and 14 are all even numbers. Any number whose last digit is 0, 2, 4, 6, or 8 is even. For contrast, see **odd number**.

EVENT In probability, an event is a set of outcomes. For example, if you toss two dice, then there are 36 possible outcomes. If A is defined to be the event where the sum of the two dice is 5, then A is a set containing four outcomes: $\{(1,4),(2,3),(3,2),(4,1)\}$. (See **probability**.)

EXISTENTIAL QUANTIFIER A backwards letter E, $\exists$, is used to represent the expression "There exists at least one. . . ," and is called the existential quantifier. For example, the sentence "There exists at least one x such that $x^2 = x$" can be written with symbols:

(1) $\exists_x(x^2 = x)$

For another example, let A_x represent the sentence "x is an American," and M_x represent the sentence "x is good at math." Then the expression

(2) $\exists_x[(A_x) \text{ AND } (M_x)]$

represents the sentence "There exists at least one x such that x is both an American and x is good at math." In more informal terms, the sentence could be written as "Some Americans are good at math."

You must be careful when you determine the negation for a sentence that uses the existential quantifier. The negation of sentence (2) is not the sentence "Some Americans are not good at math" which could be written as

(3) $\exists_x[(A_x) \text{ AND(NOT } M_x)]$

Instead, the negation of sentence (2) is the sentence "No Americans are good at math," which can be written symbolically as

(4) $\text{NOT}(\exists_x[(A_x) \text{ AND } (M_x)]$

Sentence (4) could also be written as

(5) $\forall_x[(A_x \rightarrow \text{ NOT } M_x)]$

(See **universal quantifier**.)

EXPECTATION The expectation of a discrete random variable X (written $E(X)$) is

$$E(X) = \sum_i x_i f(x_i)$$

where $f(x_i)$ is the probability function for X [that is, $f(x_i) = \text{Pr}(X = x_i)$] and the summation is taken over all possible values for X. The expectation is the average value that you would expect to see if you observed X many times. For example, if you flip a coin five times and X is the number of heads that appears, then $E(X) = 2.5$. This is what you would expect: the number of heads should be about half of the number of total flips. (Note that $E(X)$ itself does not have to be a possible value of X.)

The expectation of a continuous random variable with density function $f(x)$ is

$$E(X) = \int_{-\infty}^{\infty} x f(x) dx$$

Some properties of expectations are as follows:

$$E(A + B) = E(A) + E(B)$$
$$E(cX) = cE(X) \quad \text{if } c \text{ is a constant.}$$
$$E(AB) = E(A)E(B) + \text{Cov}(A, B)$$

(Cov(A, B) is the covariance of A and B).

The expectation is also called the *expected value*, or the *mean* of the distribution of the random variable. If the value of the summation (or the integral) used in the definition is infinite for a particular distribution, then it is said that the mean of the distribution does not exist.

EXPONENT An exponent is a number that indicates the operation of repeated multiplication. Exponents are written as little numbers or letters raised above the main line. For example:

$$3^2 = 3 \times 3 = 9$$
$$2^4 = 2 \times 2 \times 2 \times 2 = 16$$
$$10^3 = 10 \times 10 \times 10 = 1,000$$

The exponent number is also called the *power* that the base is being raised to. The second power of x (x^2) is called x squared, and the third power of x (x^3) is called x cubed.

Exponents obey these properties:

(1) Addition property: $x^a x^b = x^{a+b}$ For example:

$$4^3 \times 4^5 = (4 \times 4 \times 4)(4 \times 4 \times 4 \times 4 \times 4) = 4^8$$

(2) Subtraction property: $x^a/x^b = x^{a-b}$ For example:

$$2^6/2^2 = 2 \times 2 \times 2 \times 2 \times 2 \times 2/2 \times 2 = 2 \times 2 \times 2 \times 2 = 2^4$$

(3) Multiplication property: $(x^a)^b = x^{ab}$ For example:

$$(3^2)^3 = 3^2 \times 3^2 \times 3^2 = (3 \times 3) \times (3 \times 3) \times (3 \times 3) = 3^6$$

So far it makes sense to use only exponents that are positive integers. There are definitions that we can make, however, that will allow us to use negative exponents or fractional exponents. For negative exponents, we define:

$$x^{-a} = \frac{1}{x^a}$$

For example, $x^{-1} = 1/x$, $2^{-5} = 1/2^5 = 1/32$. This definition is consistent with the subtraction property:

$$3^{-2} = \frac{3^4}{3^6} = \frac{3 \times 3 \times 3 \times 3}{3 \times 3 \times 3 \times 3 \times 3 \times 3} = \frac{1}{3 \times 3} = \frac{1}{3^2}$$

If the exponent is zero, we define:

$$x^0 = 1$$

for all x ($x \neq 0$).

This definition seems peculiar at first, but we must have $x^0 = 1$ if the addition property of exponents is to be satisfied:

$$3^4 = 3^{4+0} = 3^4 3^0 = 3^4 \times 1$$

We define a fractional exponent to be the same as taking a root. For example, $x^{1/2} = \sqrt{x}$. By the multiplication property: $(x^{1/2})^2 = x^{2/2} = x^1$. In general: $x^{1/a} = \sqrt[a]{x}$, and $x^{a/b} = (\sqrt[b]{x})^a$. (See **root**.)

EXPONENTIAL FUNCTION An exponential function is a function of the form $f(x) = a^x$, where a is a constant known as the base. The most common exponential function is $f(x) = e^x$ (see **e**), which has the interesting property that its derivative is equal to itself. Exponential functions can be used as approximations for the rate of population growth or the growth of compound interest. The inverse function of an exponential function is the logarithm function.

EXPONENTIAL NOTATION Exponential notation provides a way of expressing very big and very small numbers on computers. A number in exponential notation is written as the product of a number from 1 to 10 and a power of 10. The letter E is used to indicate what power of 10 is needed. For example, 3.8 E 5 means 3.8×10^5. Exponential notation is the same as **scientific notation**.

EXTERIOR ANGLE (1) An exterior angle of a polygon is an angle formed by one side of the polygon and the line that is the extension of an adjacent side.

(2) When a line crosses two other lines, the four angles formed that are outside the two lines are called exterior angles. (See figure 49.)

EXTRANEOUS ROOT See **equation**.

EXTREMUM An extremum is a point where a function attains a maximum or minimum. This article will consider only functions that are continuous and differentiable. A global maximum is the point where a function attains its highest value. A local maximum is a point where the value of the function is higher than the surrounding points. Similar definitions apply to minimum points. (See figure 50.)

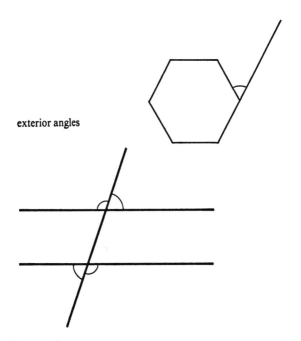

exterior angles

Figure 49

Both local maximum and local minimum points can be found by determining where the curve has a horizontal tangent, which means that the derivative is zero at that point. If the first derivative is zero and the second derivative is positive, then the curve is concave up, and the point is a minimum. For example, if $f(x) = x^2 - 10x + 7$, then the derivative is $2x - 10$, which is zero when $x = 5$. The second derivative is equal to 2, which is positive, so $(5, f(5))$ is a minimum point.

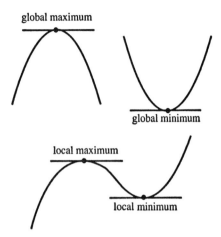

global maximum

global minimum

local maximum

local minimum

Figure 50

If the first derivative is zero and the second derivative is negative, then the curve is concave downward and the point is a maximum. For example, if $f(x) = -x^2 + 12x + 14$, then the derivative is $-2x + 12$, which is zero when $x = 6$. The second derivative is equal to -2, which is negative, so $(6, f(6))$ is a maximum point.

If the first derivative is zero and the second derivative is also zero, then the point may be a maximum, a minimum, or neither. Here are three examples:

$$f(x) = x^3; \ f'(x) = 3x^2; \ f''(x) = 6x$$

At $x = 0$ both the first and second derivative are zero, and the point $(0, f(0))$ is neither a maximum or a minimum.

$$f(x) = x^4; f'(x) = 4x^3; f''(x) = 12x^2$$

At $x = 0$ both the first and second derivative are zero, and the point $(0, f(0))$ is a minimum.

$$f(x) = -x^4; f'(x) = -4x^3; f''(x) = -12x^2$$

At $x = 0$ both the first and second derivative are zero, and the point $(0, f(0))$ is a maximum.

For the case of a function of two variables, see **second-order conditions.**

F

F-DISTRIBUTION The *F*-distribution is a continuous random variable distribution that is frequently used in statistical inference. For an example, see **analysis of variance**. There are many different *F*-distributions. Each one is identified by specifying two quantities, called the degree of freedom for the numerator (listed first) and the degree of freedom for the denominator. Table 8 at the back of the book lists some values. For example, there is a 95 percent chance that an *F*-distribution with 5 and 20 degrees of freedom will be less than 2.71. If X is a random variable with a chi-square distribution with m degrees of freedom, and Y has a chi-square distribution with n degrees of freedom that is independent of X, then this random variable:

$$\frac{X/m}{Y/n}$$

will have an *F*-distribution with m and n degrees of freedom.

FACE A polyhedron is a solid bounded by several polygons, each of which is called a face. For example, dice and all other cubes have six faces. A triangular pyramid (tetrahedron) has four faces, and a square-based pyramid has five faces.

FACTOR (1) A factor is one of two or more expressions that are multiplied together.

(2) The factors of a whole number are those whole numbers by which it can be divided with no remainder. For example, 72 has the factors 1, 2, 3, 4, 6, 8,

9, 12, 18, 24, 36, 72.

(3) To factor an expression means to express it as a product of several factors. For example, the expression $x^2 - 2x - 15$ can be factored into the following product: $(x + 3)(x - 5)$. (See **factoring**.)

FACTOR THEOREM Suppose that $P(x)$ represents a polynomial in x. The factor theorem says that, if $P(r) = 0$, then $(x - r)$ is one of the factors of $P(x)$.

FACTORIAL The factorial of a positive integer is the product of all the integers from 1 up to the integer in question. The exclamation point ("!") is used to designate factorial. For example,

$$1! = 1$$
$$2! = 2 \times 1 = 2$$
$$3! = 3 \times 2 \times 1 = 6$$
$$4! = 4 \times 3 \times 2 \times 1 = 24$$
$$5! = 5 \times 4 \times 3 \times 2 \times 1 = 120$$
$$n! = n \times (n-1) \times (n-2) \times \cdots \times 3 \times 2 \times 1$$

The factorial of zero is defined to be 1: $0!=1$.

Factorials become very big very fast. For example, 69! (read "sixty-nine factorial") is about 1.7×10^{98}. Factorials are used extensively in probability. (See **probability; permutations; combinations**.) There are $n!$ different ways of putting a group of n objects in order. For example, there are $52! = 8.1 \times 10^{67}$ ways of shuffling a deck of cards. There are 52 choices for the top card. For each choice of the top card there are 51 choices for the second card. For each of these possibilities there are 50 choices for the third card, and so on. Factorials are also used in the binomial theorem.

FACTORING Factoring is the process of splitting a complicated expression into the product of two or more simpler expressions, called factors. For example, $(x^2 - 5x + 6)$ can be split into two factors:

$$x^2 - 5x + 6 = (x - 3)(x - 2)$$

Factoring is a useful technique for solving polynomial equations and for simplifying complicated fractions. Some general tricks for factoring are:

(1) If all the terms have a common factor, then that factor can be pulled out:

$$ax^3 + bx^2 + cx = x(ax^2 + bx + c)$$

(2) The expression $x^2 + bx + c$ can be factored if you can find two numbers m and n that multiply to give c and add to give b:

$$(x + m)(x + n) = x^2 + (m + n)x + mn$$

(3) The difference of two squares can be factored:

$$a^2 - b^2 = (a - b)(a + b)$$

(4) The difference of two cubes can be factored:

$$x^3 - a^3 = (x - a)(x^2 + ax + a^2)$$

(5) The sum of two cubes can be factored:

$$x^3 + a^3 = (x + a)(x^2 - ax + a^2)$$

(See also **factor theorem**.)

FALSE "False" is one of the two truth values attached to sentences in logic. It corresponds to what we normally suppose: "false" means "not true." (See **logic; Boolean algebra**.)

FEASIBLE SOLUTION A feasible solution is a set of values for the choice variables in a linear programming problem that satisfies the constraints of the problem. (See **linear programming**.)

FERMAT Pierre de Fermat (1601 to 1665) was a French mathematician who developed number theory, worked on ideas that later became known as calculus, and corresponded with Pascal on probability theory. (See also **Fermat's last theorem**.)

FERMAT'S LAST THEOREM Fermat's last theorem states that there is no solution to the equation $a^n + b^n = c^n$ where a, b, c, and n are all positive integers, and $n > 2$. (If $n = 2$, then there are many solutions; see **Pythagorean triple**.)

The theorem acquired its name because Fermat mentioned the theorem and claimed to have discovered a proof of it, but did not have space to write it down. Nobody has ever discovered a counterexample, but it has turned out to be very difficult to prove this theorem. Over the years several proofs have been proposed, but closer analysis has revealed they have flaws. Prior to being proved, this statement should more properly be called a conjecture rather than a theorem. In 1993 Andrew Wiles proposed a proof, which started a worldwide effort to verify that the proof was correct.

FIBONACCI SEQUENCE The first two numbers of the Fibonacci sequence are 1; every other number is the sum of the two numbers that immediately precede it. Therefore, the first 14 numbers in the sequence are: 1, 1, 2, 3, 5, 8, 13, 21, 34, 55, 89, 144, 233, 377.

FIELD (1) A field is a set of elements with these properties:

—It is an Abelian group with respect to one operation called addition (with an identity element designated 0). (See **group**.)

—It is also an Abelian group with respect to another operation called multiplication.

—The distributive property holds: $a(b + c) = ab + ac$.

For example, the real numbers are an example of a field, with addition and multiplication defined in the traditional manner. The concept can also be generalized to other types of objects.

(2) See **vector field.**

FINITE Something is finite if it doesn't take forever to count or measure it. The opposite of finite is infinite, which means limitless. There is an infinite number of natural numbers. There is a finite (but very large) number of grains of sand on Palm Beach or of stars in the Milky Way galaxy.

FOCAL POINT See **ellipse; parabola; conic section.**

FOCI "Foci" is the plural of "focus." (See **focus.**)

FOCUS (1) A parabola is the set of points that are the same distance from a fixed point (the focus) and a fixed line (the directrix). The focus, or focal point, is important because starlight striking a parabolically shaped telescope mirror will be reflected back to the focus. (See **conic section; parabola; optics.**)

(2) An ellipse is the set points such that the sum

of the distances to two fixed points is a constant. The
two points are called foci (plural of focus). Planetary
orbits are shaped like ellipses, with the sun at one
focus.

FORCE A force in physics acts to cause an object
to move, or else restrains its motion. For example,
gravity is a force. A force is a vector quantity because
it has both magnitude and direction.

FOURIER Jean-Baptiste Joseph Fourier (1768 to
1830) was a French mathematician who studied differential equations of heat conduction, and developed
the concept now known as **Fourier series.**

FOURIER SERIES Any periodic function can be expressed as a series involving sines and cosines, known
as a Fourier series. Assume that units are chosen so
that the period of the function is 2π. Then:

$$f(x) = \frac{a_0}{2} + (a_1 \cos x + b_1 \sin x) + (a_2 \cos 2x + b_2 \sin 2x) + \cdots$$

$$+ (a_n \cos nx + b_n \sin nx)$$

where the coefficients are found from these integrals:

$$a_n = \frac{1}{\pi} \int_{-\pi}^{\pi} f(x) \cos nx \, dx$$

$$b_n = \frac{1}{\pi} \int_{-\pi}^{\pi} f(x) \sin nx \, dx$$

For example, consider the square wave function,
defined to be:

$$f(x) = 1 \text{ if } 0 < x < \pi, \ \ 2\pi < x < 3\pi, \ \text{ and so on}$$
$$f(x) = 0 \text{ if } -\pi < x < 0, \ \ \pi < x < 2\pi, \ \text{ and so on}$$

Set up these integrals to find the coefficients of the Fourier series (the integral only needs to be taken from 0 to π because the function is zero everywhere between $-\pi$ and 0):

$$
\begin{aligned}
b_n &= \frac{1}{\pi} \int_0^\pi \sin nx \, dx \\
&= -\frac{1}{n\pi} \cos nx \Big|_0^\pi \\
&= -\frac{1}{n\pi} [\cos(n\pi) - \cos 0] \\
&= -\frac{1}{n\pi}(-1 - 1) = \frac{2}{n\pi} \quad \text{(if } n \text{ is odd)}
\end{aligned}
$$

The remaining coefficents are zero:

$$
\begin{aligned}
a_n &= \frac{1}{\pi} \int_0^\pi \cos nx \, dx \\
&= \frac{1}{n\pi} \sin nx \Big|_0^\pi \\
&= \frac{1}{n\pi}(\sin(n\pi) - \sin 0) = 0
\end{aligned}
$$

except

$$
\begin{aligned}
a_0 &= \frac{1}{\pi} \int_0^\pi \cos 0 \, dx \\
&= \frac{1}{\pi} \int_0^\pi dx = \frac{1}{\pi}(\pi - 0) = 1
\end{aligned}
$$

Figure 51 shows how the series becomes closer to matching the square wave as more terms are added.

FRACTAL A fractal is a shape that contains an infinite amount of fine detail. That is, no matter how

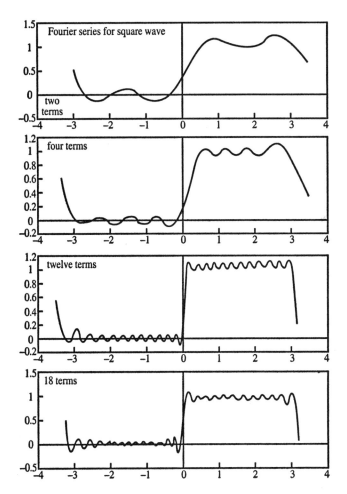

Figure 51

Figure 52   Koch snowflake generation

much it is enlarged, there is still more detail to be
revealed by enlarging it further.

Figure 52 shows how to construct a fractal called
a Koch snowflake: start with a triangle, and repeat-
edly replace every straight line by a bent line as
shown in the figure. The picture shows the result
of doing this 0, 1, 2, and 6 times. If this were done
an infinite number of times, the result would be a
fractal.

(See also **Mandelbrot set**.)

FRACTION A fraction a/b is defined by the equation

$$\frac{a}{b} \times b = a$$

The fraction a/b is the answer to the division
problem $a \div b$. The top of the fraction (a) is called
the *numerator*, and the bottom of the fraction (b) is
called the *denominator*.

Suppose that the fraction measures the amount
of pie that you have. Then the denominator tells you
how many equal slices the pie has been cut into, and
the numerator tells you how many slices you have.
The fraction $1/8$ says that the pie has been cut into
eight pieces, and you have only one of them. If you
have $8/8$, then you have eight pieces, or the whole
pie. In general, $a/a = 1$ for all a (except $a = 0$). If
$a > b$ in the fraction a/b, then you have more than a

whole pie and the value of the fraction is greater than 1. A fraction greater than 1 is sometimes called an *improper fraction*. An improper fraction can always be written as the sum of an integer and a proper fraction. For example, $\frac{10}{3} = \frac{9}{3} + \frac{1}{3} = 3 + \frac{1}{3} = 3\frac{1}{3}$.

The fraction a/b becomes larger if a becomes larger, but it becomes smaller if b becomes larger. For example, $\frac{5}{11} < \frac{6}{11}$, but $\frac{5}{11} > \frac{5}{12}$

The value of the fraction is unchanged if both the top and the bottom are multiplied by the same number: $a/b = ac/bc$ For example,

$$\frac{4}{5} = \frac{3 \times 4}{3 \times 5} = \frac{12}{15}$$

A decimal fraction, such as 0.25 (which equals $\frac{1}{4}$) is a fraction in which the part to the right of the decimal point is assumed to be the numerator of a fraction that has some power of 10 in the denominator. (See **decimal numbers**.) Decimal fractions are easier to add and compare than ordinary fractions.

A fraction is said to be in simplest form if there are no common factors between the numerator and the denominator. For example, $\frac{2}{3}$ is in simplest form because 2 and 3 have no common factors. However, $\frac{24}{30}$ is not in simplest form. To put it in simplest form, multiply both the top and the bottom by $\frac{1}{6}$.

$$\frac{\frac{1}{6} \times 24}{\frac{1}{6} \times 30} = \frac{4}{5}$$

See **least common denominator** for an example of adding two fractions.

FRUSTUM A frustum is a portion of a cone or a pyramid bounded by two parallel planes. (See fig-

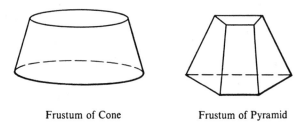

Frustum of Cone   Frustum of Pyramid

Figure 53   Frustums of cone and cylinder

ure 53.) For an application, see **surface area, figure of revolution.**

FUNCTION A function is a rule that turns each member of one set into a member of another set. The most common functions are those that turn one number into another number. For example, the function $f(x) = 3x^2 + 5$ turns 1 into 8, 2 into 17, 3 into 32, and so on. The input number to the function is called the independent variable, or argument. The set of all possible values for the independent variable is called the domain. The output number is called the dependent variable. The set of all possible values for the dependent variable is called the range.

An important property of functions is that for each value of the independent variable there is one and only one value of the dependent variable.

An inverse function does exactly the opposite of the original function. If you put x into the original function and get out y, then, if you put y into the inverse function, you will get out x. In order for a function to have an inverse, it must be one-to-one; that is, there must be one and only one input number for each output number. (It is possible for a function to have two input numbers leading to the same output number, but such a function will not have an inverse.) The range of the inverse function is the same as the domain of the original function and vice versa. For example, the natural logarithm function is the inverse of the exponential function e^x.

FUNDAMENTAL THEOREM OF ALGEBRA

The fundamental theorem of algebra says that an nth-degree polynomial equation has at least one root among the complex numbers. It has exactly n roots when you include complex roots and you realize that a root may occur more than once. (See **polynomial**.)

FUNDAMENTAL THEOREM OF ARITHMETIC

The fundamental theorem of arithmetic says that any natural number can be expressed as a unique product of prime numbers. (See **prime factors**.)

FUNDAMENTAL THEOREM OF CALCULUS

The fundamental theorem of calculus says that

$$\lim_{n\to\infty,\Delta x\to 0} \sum_{i=1}^{n} f(x_i)\Delta x = \int_a^b f(x)dx = F(b) - F(a)$$

where $\Delta x = (b-a)/n$, x_i is a number in the interval from $a + (i-1)\Delta x$ to $a + i\Delta x$, and $dF(x)/dx = f(x)$. The theorem tells how to find the area under a curve by taking an integral. (See **calculus, definite integral**.)

G

GALOIS Evariste Galois (1811 to 1832) was a French
mathematician who made crucial contributions to
group theory and applied this to the study of the
solvability of polynomial equations.

GAUSS Carl Friedrich Gauss (1777 to 1855) was a
German mathematician and astronomer who studied errors of measurement (so the normal curve is
sometimes called the Gaussian error curve); developed a way to construct a 17-sided regular polygon
with geometric construction; developed a law that
says the electric flux over a closed surface is proportional to the charge inside the surface (this law is
now included as one of **Maxwell's equations**); and
studied the theory of complex numbers.

GAUSS-JORDAN ELIMINATION Gauss-Jordan
elimination is a method for solving a system of linear equations. The method involves transforming the
system so that the last equation contains only one
variable, the next-to-last equation contains only two
variables, and so on. The system is easy to solve
when it is in that form. For example, to solve this
system:

$$2x - 3y + z = 5$$
$$6x + y - 5z = 51$$
$$4x + 14y - 8z = 100$$

eliminate the term with x from the last two equations. To do this, subtract twice the first equation
from the last equation to obtain a new last equation,

and subtract three times the first equation from the
second equation to obtain a new second equation.
The system then looks like this:

$$2x - 3y + z = 5$$
$$10y - 8z = 36$$
$$20y - 10z = 90$$

Now, to eliminate the term with y from the last
equation, subtract twice the second equation from
the last equation. Here is the new system:

$$2x - 3y + z = 5$$
$$10y - 8z = 36$$
$$6z = 18$$

Solve the last equation for z (solution: $z = 3$).
Then insert this value for z into the second equation
to solve for y (solution: $y = 6$). Finally, insert the
values for z and y into the first equation to solve for
x (solution: $x = 10$).

GEOMETRIC CONSTRUCTION Geometric construction is the process of drawing geometric figures
using only two instruments: a straightedge and a
compass. Figure 54 shows how to bisect an angle
with geometric construction. First, put the point
of the compass at the vertex of the angle, and then
mark off arcs 1 and 2 (each equal distance from the
vertex). Then, put the point of the compass at the
point where arc 1 crosses the side of the angle, and
then mark off arc 3. Move the point to arc 2, and
then mark off arc 4 (making sure that the distance is

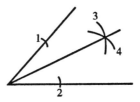

Figure 54   Bisecting an angle with geometric construction

the same as it was from arc 1 to arc 3). To bisect the angle, simply draw the line connecting the vertex of the angle to the point where arcs 3 and 4 cross.

Classical geometers sought a similar way of trisecting an angle with geometric construction, but that has since been proved to be impossible.

GEOMETRIC DISTRIBUTION Consider a random experiment where the probablity of success on each trial is p. You will keep conducting the experiment until you see the first success; let X be the number of failures that occur before the first success. (Assume that each trial is independent of the others.) Then X is a discrete random variable with the geometric distribution. Its probability function is:

$$Pr(X = i) = p(1 - p)^i$$

The expectation of X is $(1 - p)/p$, and the variance is $(1 - p)/p^2$. For exampl , if you are trying to roll a 6 on one die, then $p = 1/6$, and you can expect to roll 5 non-sixes before rolling a 6. For comparison, see **binomial distribution**.

GEOMETRIC MEAN The geometric mean of a group of n numbers $(a_1, a_2, a_3, \ldots a_n)$ is equal to

$$(a_1 \times a_2 \times a_3 \times \ldots \times a_n)^{1/n}$$

For example, the geometric mean of 4 and 9 is $\sqrt{4 \times 9}$ = 6. For contrast, see **arithmetic mean**.

GEOMETRIC SEQUENCE A geometric sequence is a sequence of numbers of the form

$$a, ar, ar^2, ar^3, \ldots ar^{n-1}$$

The ratio between any two consecutive terms is a constant.

GEOMETRIC SERIES A geometric series is a sum of a geometric sequence:

$$S = a + ar + ar^2 + ar^3 + ar^4 + \ldots + ar^{n-1}$$

In a geometric series the ratio of any two consecutive terms is a constant (in this case r). The sum of the n terms of the geometric series above is

$$\sum_{i=0}^{n-1} ar^i = \frac{a(r^n - 1)}{r - 1}$$

To show this, multiply the series by $(1 - r)$:

$$(a + ar + ar^2 + ar^3 + ar^4 + \ldots + ar^{n-1})(1 - r) =$$
$$a + ar + ar^2 + ar^3 + ar^4 + \ldots + ar^{n-1}$$
$$-ar - ar^2 - ar^3 - ar^4 - \ldots - ar^{n-1} - ar^n$$
$$= a - ar^n$$

Therefore,

$$S = a + ar + ar^2 + ar^3 + ar^4 + \ldots + ar^{n-1} = \frac{a - ar^n}{1 - r}$$

which can be rewritten in the form given above.

For example:

$$2 + 4 + 8 + 16 + 32 + 64 = \frac{(2)(2^6 - 1)}{2 - 1} = 126$$

If n approaches infinity, then the summation will also go to infinity if $|r| > 1$ However, if $-1 < r < 1$, then r^n approaches zero as n approaches infinity, so the expression for the sum of the terms becomes:

$$\sum_{i=0}^{\infty} ar^i = \frac{a}{1 - r}$$

For example:

$$1 + \frac{1}{2} + \frac{1}{4} + \frac{1}{8} + \frac{1}{16} + \frac{1}{32} + \ldots = \frac{1}{1 - \frac{1}{2}} = 2$$

GEOMETRY Geometry is the study of shape and size. The geometry of our everyday world is based on the work of Euclid, who lived about 300 B.C. Euclidian geometry has a rigorously developed logical structure. Three basic undefined terms are point, line, and plane. A point is like a tiny dot: it has zero height, zero width, and zero thickness. A line goes off straight in both directions. A plane is a flat surface, like a tabletop, extending off to infinity. We cannot see any of these idealized objects, but we can imagine them and draw pictures to represent them. Euclid developed some basic postulates and then proved theorems based on these. Examples of postulates used in modern versions of Euclidian geometry are "Two distinct points are contained in one and only one line" and "Three distinct points not on the same line are contained in one and only one plane."

The geometry of flat figures is called plane geometry, because a flat figure is contained in a plane. The geometry of figures in three dimensional space is called solid geometry.

Other types of geometries (called non-Euclidian geometries) have been developed, which make different assumptions about the nature of parallel lines. Although these geometries do not match our intuitive concept of what space is like, they have been useful in developing general relativity theory and in other areas of math.

GÖDEL Kurt Gödel (1906 to 1978) was an Austrian born U.S. mathematician who developed **Gödel's incompleteness theorem.**

GÖDEL'S INCOMPLETENESS THEOREM This theorem states that a rigid logical system will contain true propositions that cannot be proved to be true. Therefore, no logical system can be complete in the sense of being able to provide formal proofs for all true theorems.

GRADIENT The gradient of a multivariable function is a vector consisting of the partial derivatives of that function. If $f(x, y, z)$ is a function of three variables, then the gradient of f, written as ∇f, is the vector

$$\left(\frac{\partial f}{\partial x}, \frac{\partial f}{\partial y}, \frac{\partial f}{\partial z} \right)$$

For example, if

$$f(x, y, z) = x^a y^b z^c$$

then the gradient is the vector

$$[(ax^{a-1}y^b z^c), (bx^a y^{b-1} z^c), (cx^a y^b z^{c-1})]$$

GRAPH 146

If the gradient is evaluated at a particular point (x_1, y_1, z_1), then the gradient points in the direction of the greatest increase of the function starting at that point. If the gradient is equal to the zero vector at a particular point, then that point is a critical point that might be a local maximum or minimum. (See **extremum; second-order conditions.**)

GRAPH The graph of an equation is the set of points that make the equation true. By drawing a picture of the graph it is possible to visualize an algebraic equation. For example, the set of points that make the equation $x^2 + y^2 = r^2$ true is a circle.

GREAT CIRCLE A great circle is a circle that is formed by the intersection of a sphere and a plane passing through the center. (See **sphere; spherical trigonometry.**)

GREATEST COMMON FACTOR The greatest common factor of two natural numbers a and b is the largest natural number that divides both a and b evenly (that is, with no remainder). For example, the greatest common factor of 15 and 28 is 1. The greatest common factor of 60 and 84 is 12. (See **Euclid's algorithm.**)

GREEN'S THEOREM Let $f(x, y) = [f_x(x, y), f_y(x, y)]$ be a two-dimensional vector field, and let L be a closed path in the x, y plane. Green's theorem states that the line integral of f around this path is equal to the following integral over the interior of the path L:

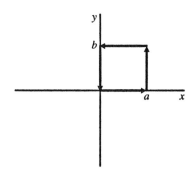

Figure 55 Green's theorem for rectangular path

$$\int_{path\ L} \mathbf{f}(x,y)\mathbf{dL} = \int\int_{interior\ of\ L} \left[\frac{\partial f_y}{\partial x} - \frac{\partial f_x}{\partial y} \right] dx\ dy$$

The amazing part of this theorem is that it works for any vector field $\mathbf{f}$ and any path L. The following will show that it works for the rectangular path shown in figure 55; this result can be generalized to an arbitrary path.

Start with the double integral over the interior:

$$\int_{y=0}^{y=b} \int_{x=0}^{x=a} \left[\frac{\partial f_y}{\partial x} - \frac{\partial f_x}{\partial y} \right] dx\ dy$$

$$= \int_{y=0}^{y=b} \int_{x=0}^{x=a} \left(\frac{\partial f_y}{\partial x} \right) dx\ dy - \int_{y=0}^{y=b} \int_{x=0}^{x=a} \left(\frac{\partial f_x}{\partial y} \right) dx\ dy$$

$$= \int_{y=0}^{y=b} [f_y|_{x=0}^{x=a}]\ dy - \int_{x=0}^{x=a} \int_{y=0}^{y=b} \frac{\partial f_x}{\partial y}\ dy\ dx$$

$$= \int_0^b [f_y(a,y) - f_y(0,y)]dy - \int_{x=0}^{x=a} [f_x|_{y=0}^{y=b}]dx$$

This can be rearranged into these four integrals:

$$\int_0^a f_x(x,0)dx + \int_0^b f_y(a,y)dy +$$
$$\int_a^0 f_x(x,b)dx + \int_b^0 f_y(0,y)dy$$

When combined, these four integrals give the four pieces of the line integral around the rectangular path.

For a generalization of this result, see **Stokes's theorem.** For an application, see **Maxwell's equations.** For background, see **line integral.**

GROUP A group is a set of elements for which an operation (call it ∘) is defined that meets these properties:

(1) If a and b are in the set, then $a \circ b$ is also in the set.

(2) The associative property holds:
$a \circ (b \circ c) = (a \circ b) \circ c$

(3) There is an identity element I such that
$a \circ I = a$

(4) Each element (a) has an inverse (a^{-1}) such that $a \circ a^{-1} = I$.

If the operation is also commutative (that is, $a \circ b = b \circ a$, then the group is called an Abelian group.

For example, the real numbers form an Abelian group with respect to addition, and the nonzero real numbers form an Abelian group with respect to multiplication. The theory of groups can be applied to many sets other than numbers, and to operations other than conventional multiplication.

(See also **field.**)

H

HALF PLANE A half plane is the set of all points in a plane that lie on one side of a line.

HARMONIC SEQUENCE A sequence of numbers is a harmonic sequence if the reciprocals of the terms form an arithmetic sequence. The general form of a harmonic sequence is

$$\frac{1}{a}, \frac{1}{a+d}, \frac{1}{a+2d}, \frac{1}{a+3d}, \cdots \frac{1}{a+(n-1)d}$$

HEPTAGON A heptagon is a polygon with seven sides.

HERO'S FORMULA Hero's formula tells how to find the area of a triangle if you know the length of the sides. Let a, b, and c be the lengths of the sides, and let $s = (a+b+c)/2$. Then the area of the triangle is given by the formula

$$\sqrt{s(s-a)(s-b)(s-c)}$$

HEXADECIMAL NUMBER A hexadecimal number is a number written in base 16. A hexadecimal system consists of 16 possible digits. The digits from 0 to 9 are the same as they are in the decimal system. The letter A is used to represent 10; B = 11; C = 12; D = 13; E = 14; and F = 15. For example, the number A4C2 in hexadecimal means

$$10 \times 16^3 + 4 \times 16^2 + 12 \times 16^1 + 2 \times 16^0 = 42,178$$

HEXAGON A hexagon is a six-sided polygon. The sum of the angles in a hexagon is 720°. Regular

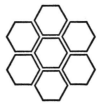

Figure 56 Hexagons

hexagons have six equal sides and six equal angles
of 120°. Honeycombs are shaped like hexagons for
a good reason. With a fixed perimeter, the area of
a polygon increases as the number of sides increases. If you have a fixed amount of fencing, you will
have more area if you build a square rather than a
triangle. A pentagon would be even better, and a
circle would be best of all. There is one disadvan-
tage to adding more sides, though. If a polygon has
too many sides, you can't pack several of those polygons together without wasting a lot of space. You
can't pack circles tightly, or even octagons. You can
pack hexagons, though. Hexagons make a nice com-
promise: they have more area for a fixed perimeter
than any other polygon that can be packed together
tightly with others of the same type. (See figure 56.)

HEXAHEDRON A hexahedron is a polyhedron with
six faces. A regular hexahedron is better known as a
cube.

HYPERBOLA A hyperbola is the set of all points
in a plane such that the difference between the distances to two fixed points is a constant. A hyperbola

has two branches that are mirror images of each other. Each branch looks like a misshaped parabola. The general equation for a hyperbola with center at the origin is

$$\frac{x^2}{a^2} - \frac{y^2}{b^2} = 1$$

The meaning of a and b is shown in figure 57. The two diagonal lines are called *asymptotes*. The farther you are from the origin, the closer each part of the curve approaches its respective asymptote line. However, the curve never actually touches the lines.

HYPERBOLIC FUNCTIONS The hyperbolic functions are a set of functions defined as follows:

hyperbolic cosine: $\cosh x = \frac{1}{2}(e^x + e^{-x})$

hyperbolic sine: $\sinh x = \frac{1}{2}(e^x - e^{-x})$

hyperbolic tangent: $\tanh x = \frac{\sinh x}{\cosh x}$

For an example of an application, see **catenary**.

HYPERGEOMETRIC DISTRIBUTION The hypergeometric distribution is a discrete random variable distribution that applies when you are selecting a sample without replacement from a population. Suppose that the population contains M "desirable" objects and $N - M$ "undesirable" objects. Select n objects from the population at random without replacement (in other words, once an object has been selected, you will not return it to the population and therefore it cannot be selected again). Let X be the number of desirable objects in your sample. Then X is a discrete random variable

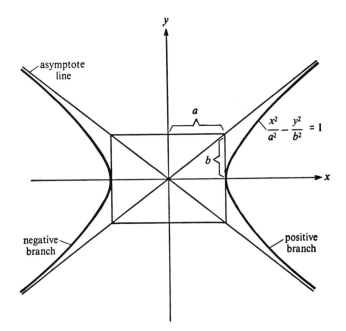

Figure 57 Hyperbola

with the hypergeometric distribution. Its probability function is given by this formula:

$$Pr(X = i) = \frac{\binom{M}{i} \times \binom{N - M}{n - i}}{\binom{N}{n}}$$

The symbols in the parentheses are all examples of the binomial coefficient. For example:

$$\binom{N}{n} = \frac{N!}{(N - n)!n!}$$

(See **combinations; binomial theorem**.)

The expected value of X is equal to nM/N. Here is the intuition for this result. If you have $M = 600$ blue marbles in a jar with a total of $N = 1000$ marbles, and you randomly select $n = 100$ marbles from the jar, you would expect to choose about 60 blue marbles.

The variance of X is $np(1 - p)(N - n)/(N - 1)$, where $p = M/N$, the proportion of desirable objects in the population.

HYPERPLANE A hyperplane is the generalization of the concept of a plane to higher dimensional space. A plane (in 3 dimensions) can be defined by an equation of the form $ax + by + cz = d$, where a, b, c and d are known constants. A hyperplane of dimension n can be defined by an equation of the form:

$$a_1 x_1 + a_2 x_2 + \cdots + a_n x_n = a_0$$

where a_0 to a_n are known constants.

HYPOTENUSE The hypotenuse is the side in a right triangle that is opposite the right angle. It is the longest of the three sides in the triangle.

(See **Pythagorean theorem**.)

HYPOTHESIS A hypothesis is a proposition that is being investigated; it has yet to be proved. (See **hypothesis testing**.)

HYPOTHESIS TESTING A situation often arises in which a researcher needs to test a hypothesis about the nature of the world. Frequently it is necessary to use a statistical technique known as hypothesis testing for this purpose.

The hypothesis that is being tested is termed the *null hypothesis*. The other possible hypothesis, which says "The null hypothesis is wrong," is called the *alternative hypothesis*. Here are some examples of possible null hypotheses:

"There is no significant difference in effectiveness between Brand X cold medicine and Brand Z medicine."

"On average, the favorite colors for Democrats are the same as the favorite colors for Republicans."

"The average reading ability of fourth graders who watch less than 10 hours of television per week is above that of fourth graders who watch more than 10 hours of television."

The term "null hypothesis" is used because the hypothesis that is being tested is often of the form "There is no relation between two quantities," as in the first example above. However, the term "null hypothesis" is used also in other cases whether or not it is a "no-effect" type of hypothesis.

In many practical situations it is not possible to determine with certainty whether the null hypothesis is true or false. The best that can be done is to collect evidence and then decide whether the null hypothesis should be accepted or rejected. There is always a possibility that the researcher will choose incorrectly, since the truth is not known conclusively. A situation in which the null hypothesis has been rejected, but is actually true, is referred to as a type 1 error. The opposite type of error, called a type 2 error, occurs when the null hypothesis has been accepted, but is actually false. A good testing procedure is designed so that the chance of committing either of these errors is small. However, it often works out that a test procedure with a smaller probability of leading to a type 1 error will also have a larger probability of resulting in a type 2 error. Therefore, no single testing procedure is guaranteed to be best. It is customary in statistics to design a testing procedure such that the probability of a type 1 error is less than a specified value (often 5 percent or 1 percent). The probability of committing a type 1 error is called the *level of significance* of the test. Therefore, if a test has been conducted at the 5 percent level of significance, this means that the test has been designed so that there is a 5 percent chance of a type 1 error.

The normal procedure in hypothesis testing is to calculate a quantity called a *test statistic*, whose value depends on the values that are observed in the sample. The test statistic is designed so that if the null hypothesis is true, then the test statistic value will be a random variable that comes from a known distribution, such as the standard normal distribution or a *t* distribution. After the value of the test statistic has been calculated, that value is compared with the values that would be expected from the known distribution. If the observed test statistic value might plausibly have come from the indicated distribution, then the null hypothesis is accepted. However, if it is unlikely that the observed value could have resulted from that distribution, then the null hypothesis is rejected.

Suppose that we are conducting a test based on a test statistic Z, which will have a standard normal distribution if the null hypothesis is true. There

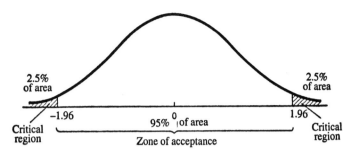

Figure 58 Hypothesis testing

is a 95 percent chance that the value of a random variable with a standard normal distribution will be between 1.96 and −1.96. Therefore, we will design the test so that the null hypothesis will be accepted if the calculated value of Z falls between −1.96 and 1.96, since these are plausible values. However, if the value of Z is less than −1.96 or greater than 1.96, we will reject the hypothesis because the value of a random variable with a standard normal distribution is unlikely to fall outside the −1.96 to 1.96 range. The range of values for the test statistic where the null hypothesis is rejected is known as the *rejection region* or *critical region.* In this case the critical region consists of two parts. (The two regions at the ends of the distribution are called the tails of the distribution.) Notice that there still is a 5 percent chance of committing a type 1 error. If the null hypothesis is true, then Z will have a standard normal distribution, and there is a 5 percent chance that the value of Z will be greater than 1.96 or less than −1.96. (See figure 58.)

Here is an example of a hypothesis testing problem involving coins. Suppose we wish to test whether a particular coin is fair (that is, equally likely to come up heads or tails). Our null hypothesis is "The probability of heads is .5." The alternative hypothesis is "The probability of heads is not .5." To conduct our test, we will flip the coin 10,000 times. Let X be the number of heads that occurs; X is a random variable. If the null hypothesis is true, then X has a binomial distribution with $n = 10,000, p = .5, E(X) = np = 5,000, Var(X) = np(1-p) = 2,500$, and standard deviation $= 50$. Because of the central limit theorem, X can be approximated by a normal distribution with mean 5,000 and standard deviation 50. We define a new random variable Z as follows: $Z = (X - 5000)/50$. Now Z will have a standard normal distribution. If the calculated value of Z is between -1.96 and 1.96, we will accept the null hypothesis that the coin is fair; otherwise we will reject the hypothesis. For example, if we observe 5063 heads, then $X = 5063$, $Z = 1.26$, and we will accept the null hypothesis. On the other hand, if we observe 5104 heads, then $X = 5104$, $Z = 2.08$, and we will reject the null hypothesis because the observed value of Z falls in the critical region.

For other examples of hypothesis testing, see **chi-square test** and **analysis of variance.**

I

i The symbol *i* is the basic unit for imaginary numbers, and is defined by the equation $i^2 = -1$. (See **imaginary number**.)

ICOSAHEDRON An icosahedron is a polyhedron with 20 faces. (See **polyhedron**.) (See figure 59.)

IDENTITY An identity is an equation that is true for every possible value of the unknowns. For example, the equation $4x = x + x + x + x$ is an identity, but $2x + 3 = 15$ is not.

IDENTITY ELEMENT If ∘ stands for an operation (such as addition), then the identity element (called *I*) for the operation ∘ is the number such that $I \circ a = a$, for all *a*. For example, zero is the identity element for addition, because $0 + a = a$, for all *a*. One is the identity element for multiplication, because $1 \times a = a$, for all *a*.

IDENTITY MATRIX An identity matrix is a square matrix with ones along the diagonal and zeros everywhere else. For example:

Figure 59 Icosahedron

$(2 \times 2$ identity):

$$\begin{pmatrix} 1 & 0 \\ 0 & 1 \end{pmatrix}$$

$(3 \times 3$ identity):

$$\begin{pmatrix} 1 & 0 & 0 \\ 0 & 1 & 0 \\ 0 & 0 & 1 \end{pmatrix}$$

$(4 \times 4$ identity):

$$\begin{pmatrix} 1 & 0 & 0 & 0 \\ 0 & 1 & 0 & 0 \\ 0 & 0 & 1 & 0 \\ 0 & 0 & 0 & 1 \end{pmatrix}$$

The letter **I** is used to represent an identity matrix. An identity matrix satisfies the property that **IA** = **A** for any matrix for which **IA** exists. For example:

$$\begin{pmatrix} 1 & 0 & 0 \\ 0 & 1 & 0 \\ 0 & 0 & 1 \end{pmatrix} \begin{pmatrix} 11 & 12 & 13 \\ 21 & 22 & 23 \\ 31 & 32 & 33 \end{pmatrix} = \begin{pmatrix} 11 & 12 & 13 \\ 21 & 22 & 23 \\ 31 & 32 & 33 \end{pmatrix}$$

If the result when multiplying two square matrices is the identity matrix, then each matrix is called the inverse matrix for the other. (See **inverse matrix.**)

IF The word "IF" in logic is used in conditional statements of the form "IF p, THEN q" $(p \rightarrow q)$. (See **conditional statement.**)

IMAGE The image of a point is the point that results after the original point has been subjected to a transformation. For an example of a transformation, see **reflection**.

IMAGINARY NUMBER An imaginary number is of the form ni, where n is a real number that is being multiplied by the imaginary unit i, and i is defined by the equation $i^2 = -1$. Since the product of any two real numbers with the same sign will be positive (or zero), there is no way that you can find any real number that, when multiplied by itself, will give you a negative number. Therefore, the imaginary numbers need to be introduced to provide solutions for equations that require taking the square roots of negative numbers.

Imaginary numbers are needed to describe certain equations in some branches of physics, such as quantum mechanics. However, any measurable quantity, such as energy, momentum, or length, will always be represented by a real number.

The square root of any negative number can be expressed as a pure imaginary number:

$$\sqrt{(-10)} = \sqrt{(-1)(10)} = \sqrt{-1}\sqrt{10} = i\sqrt{10}$$

An interesting cyclic property occurs when i is raised to powers:

$$
\begin{array}{lll}
i^0 = 1 & i^4 = 1 & i^8 = 1 \\
i^1 = i & i^5 = i & i^9 = i \\
i^2 = -1 & i^6 = -1 & i^{10} = -1 \\
i^3 = -i & i^7 = -i & i^{11} = -i
\end{array}
$$

A **complex number** is formed by the addition of a pure imaginary number and a real number. The general form of a complex number is $a + bi$, where a and b are both real numbers.

IMPLICATION An implication is a statement of this form; "$A \rightarrow B$" ("A implies B"). (See **conditional statement**.)

IMPLICIT DIFFERENTIATION Implicit differentiation provides a method for finding derivatives if the relationship between two variables is not expressed as an explicit function. For example, consider the equation $x^2 + y^2 = r^2$, which describes a circle of radius r centered at the origin. This equation defines a relationship between x and y, but it does not express that relationship as an explicit function. To find the derivative dy/dx, take the derivative of both sides of the equation with respect to x:

$$\frac{d}{dx}(x^2 + y^2) = \frac{d}{dx}(r^2)$$

$$\frac{d(x^2)}{dx} + \frac{d(y^2)}{dx} = \frac{d(r^2)}{dx}$$

Assume that r is a constant; then $d(r^2)/dx$ is zero. Use the chain rule to find the two derivatives on the left:

$$2x + 2y\frac{dy}{dx} = 0$$

Now solve for dy/dx:

$$\frac{dy}{dx} = -\frac{x}{y}$$

For another example, suppose that $y = a^x$. Take the logarithm of both sides:

$$\ln y = x \ln a$$

Now y is no longer written as an explicit function of x, but you can again use implicit differentiation:

$$\frac{d}{dx}(\ln y) = \frac{d}{dx}(x \ln a)$$

Assume that a is a constant:

$$\frac{d}{dx}(\ln y) = \ln a$$

Use the chain rule on the left-hand side:

$$\frac{1}{y}\frac{dy}{dx} = \ln a$$

and then solve for dy/dx:

$$\frac{dy}{dx} = y \ln a = a^x \ln a$$

IMPROPER FRACTION An improper fraction is a fraction with a numerator that is greater than the denominator: $\frac{7}{4}$, for example. An improper fraction can be written as the sum of a whole number and a proper fraction. For example, $\frac{7}{4} = 1 + \frac{3}{4} = 1\frac{3}{4}$. For contrast, see **proper fraction**.

INCENTER The incenter of a triangle is the center of the circle inscribed inside the triangle. It is the intersection of the three angle bisectors of the triangle. (See **incircle**.)

INCIRCLE The incircle of a triangle is the circle that can be inscribed within the triangle. (See figure 60.) For contrast, see **circumcircle**.

INCONSISTENT EQUATIONS Two equations are inconsistent if they contradict each other and therefore cannot be solved simultaneously. For example, $2x = 4$ and $3x = 9$ are inconsistent. (See **simultaneous equations**.)

Figure 60 Incircle

INCREASING FUNCTION A function $f(x)$ is an increasing function if $f(a) > f(b)$ when $a > b$.

INCREMENT In mathematics, the word "increment" means "change in." An increment in a variable x is usually symbolized as Δx.

INDEFINITE INTEGRAL The indefinite integral of a function f is symbolized as follows:

$$\int f(x)dx = F(x) + C$$

where $\int$ is the integral sign, and F is an antiderivative function for f [that is, $dF/dx = f(x)$]. C is called the arbitrary constant of integration. Since the derivative of a constant is equal to zero, it is possible to add any constant to a function without changing its derivative. That is the reason why this type of integral is called an indefinite integral. For example, suppose that a car is driven at a constant speed of 55 miles per hour. Then its position at time t will be given by the indefinite integral

$$\int 55dt = 55t + C$$

Because of the arbitrary constant, we do not know the exact value of the position. We know that the car has been traveling 55 miles per hour, but we cannot figure out its position unless we also know where it started from. If the car started at milepost 25 at time zero, we can solve for the value of the arbitrary constant, and then we will know that the position of the car at time t is given by the function $55t + 25$.

In general, it is possible to solve for the arbitrary constant of integration if we are given an initial condition.

(See also **integral; definite integral.**)

INDEPENDENT EVENTS Two events are independent if they do not affect each other. For example, the probability that a new baby will be a girl is not affected by the fact that a previous baby was a girl. Therefore, these two events are independent. If A and B are two independent events, the conditional probability that A will occur, given that B has occurred, is just the same as the unconditional probability that A will occur:

$$\Pr(A|B) = \Pr(A)$$

(See **conditional probability.**)

Also, if A and B are independent, the probability that both A and B will occur is equal to the probability of A times the probability of B:

$$\Pr(A \text{ AND } B) = \Pr(A) \times \Pr(B)$$

For example, suppose the probability that the primary navigation system on a spacecraft will fail is .01, the probability that the backup navigation system will fail is .05, and these two events are independent. In other words, the probability that the backup system will fail is not affected by whether or not the primary system has failed. Then the probability that both systems will fail is $.01 \times .05 = .0005$. Therefore, the probability that both systems will fail is much smaller than the probability that either of the individual systems will fail. This result would not be true, however, if these two events were not independent. If the probability that the backup system will fail rises if the primary system has failed, then the spacecraft could be in trouble.

INDEPENDENT VARIABLE The independent variable is the input number to a function. In the equation $y = f(x)$, x is the independent variable and y is the dependent variable. (See **function**.)

INDEX The index of a radical is the little number that tells what root is to be taken. For example, in the expression $\sqrt[3]{64} = 4$, the number 3 is the index of the radical. It means to take the cube root of 64. If no index is specified, then the square root is assumed: $\sqrt[2]{36} = \sqrt{36} = 6$.

INDIRECT PROOF The method of indirect proof begins by assuming that a theorem is false, and then proceeds to show that a contradiction results. Therefore, the theorem must be true. For an example, see **irrational number**.

INDUCTION Induction is the process of reasoning from a particular circumstance to a general conclusion. (See **mathematical induction**.)

INEQUALITY An inequality is a statement of this form: "x is less than y," written as $x < y$, or "x is greater than y," written as $x > y$. The arrow in the inequality sign always points to the smaller number. Inequalities containing numbers will either be true (such as $8 > 7$), or false (such as $4 < 3$). Inequalities containing variables (such as $x < 3$) will usually be true for some values of the variable.

The symbol $\leq$ means "is less than or equal to," and the symbol $\geq$ means "is greater than or equal to."

A true inequality will still be true if you add or subtract the same quantity from both sides of the inequality. The inequality will still be true if both sides are multiplied by the same positive number, but if you multiply by a negative number you must reverse the inequality:

$$
\begin{array}{c|c}
4 > 3 & 4 > 3 \\
2 \times 4 > 2 \times 3 & -2 \times 4 < -2 \times 3 \\
8 > 6 & -8 < -6
\end{array}
$$

(See also **system of inequalities**.)

INFINITE SERIES An infinite series is the sum of an infinite number of terms. In some cases the series may have a finite sum. (See **geometric series**.)

INFINITESIMAL An infinitesimal is a variable quantity that approaches very close to zero. In calculus Δx is usually used to represent an infinitesimal change in x. Infinitesimals play an important role in the study of limits.

point of
inflection

Figure 61

INFINITY The symbol "∞" (infinity) represents a
limitless quantity. It would take you forever to count
an infinite number of objects. There is an infinite
number of numbers. As x goes to zero, the quantity
$1/x$ goes to infinity. (However, that does not mean
that there is a number called ∞ such that $1/0 = \infty$.)
The opposite of "infinite" is finite.

INFLECTION POINT An inflection point on a
curve is a point such that the curve is oriented
concave-upward on one side of the point and concave-
downward on the other side of the point. (See figure 61.) If the curve represents the function $y =
f(x)$, then the second derivative d^2y/dx^2 is equal to
zero at the inflection point.

INSCRIBED (1) An inscribed polygon is a polygon
placed inside a circle so that each vertex of the polygon touches the circle. For an example, see **pi**.
    (2) An inscribed circle of a polygon is a circle lo-
cated inside a polygon, with each side of the polygon
being tangent to the circle. For an example, see **incircle**. A circle can be inscribed in any triangle or

regular polygon. There are many polygons, such as a rectangle, where it is not possible to inscribe a circle that touches each side.

INTEGERS The set of integers contains zero, the natural numbers, and the negatives of all the natural numbers:

$$\ldots, -6, -5, -4, -3, -2, -1, 0, 1, 2, 3, 4, 5, 6, \ldots$$

An integer is a real number that does not include a fractional part. The natural numbers are also called the positive integers, and the integers smaller than zero are called the negative integers.

INTEGRAL The process of finding an integral (called *integration*) is the reverse process of finding a derivative. The *indefinite integral* of a function $f(x)$ is a function $F(x) + C$ such that the derivative of $F(x)$ is equal to $f(x)$, and C is an arbitrary constant. The indefinite integral is written with the integral sign:

$$\int f(x)dx = F(x) + C$$

(See **calculus; derivative; indefinite integral.**) Here is a table of integrals of some functions:

Perfect Integral Rule

$$\int dx = x + C$$

in other words, if $f(x) = 1$, then $F(x) = x$

Sum Rule

$$\int [f(x) + g(x)]dx = \int f(x)dx + \int g(x)dx$$

Multiplication by a constant

$$\int a f(x)dx = a \int f(x)dx$$

(if a is a constant)

Power Rule

$$\int x^n dx = \frac{1}{n+1}x^{n+1} + C \quad (\text{if } n \neq -1)$$

$$\int x^{-1}dx = \ln|x| + C$$

(The above rules make it possible to find the integral of any polynomial function.)

Trigonometric Integrals

$$\int \sin x \, dx = -\cos x + C$$

$$\int \cos x \, dx = \sin x + C$$

$$\int \tan x \, dx = \ln|\sec x| + C$$

$$\int \sec x \, dx = \ln|\sec x + \tan x| + C$$

For more information on integration methods, see **integration by trigonometric substitution** and **integration by parts**. Also, table 9 at the back of the book lists many common integrals.

Integrals can also be used to find the area under curves and other quantities. (See **definite integral; surface area, figure of revolution; volume, figure of revolution; arc length; centroid.**)

INTEGRAND The integrand is a function that is to be integrated. In the expression $\int f(x)dx$, the function $f(x)$ is the integrand. (See **integral**.)

INTEGRATION Integration is the process of finding an integral. (See **integral**.)

INTEGRATION BY PARTS Integration by parts is a method for solving some difficult integrals that is based on a formula found by reversing the product rule for derivatives:

$$\int u\ dv = uv - \int v\ du$$

The key to making this method work is to define u and dv in a fashion such that the integral $\int v\ du$ will be easier to solve than the original integral ($\int u\ dv$).

For example, $\int \ln x dx$ can be integrated by defining $u = \ln x$, $dv = dx$. Then:

$$du = \frac{1}{x}dx, \quad v = x$$

$$\begin{aligned}\int \ln x dx &= x \ln x - \int x\frac{1}{x}dx \\ &= x \ln x - x + C\end{aligned}$$

For another example, to solve $\int x \cos x dx$, let $u = x$ and $dv = \cos x dx$. Then $du = dx$, and $v = \sin x$.

$$\begin{aligned}\int x \cos x dx &= x \sin x - \int \sin x dx \\ &= x \sin x + \cos x + C\end{aligned}$$

Integration by parts is sometimes a trial and error process, as it is not obvious in advance which integrals the method will work for, and it is not always clear the best way to make the definitions u and dv.

**INTEGRATION BY TRIGONOMETRIC SUB-
STITUTION** Some integrals involving expressions
of the form $(1+x^2)$ or $(1-x^2)$ can be solved by making trigonometric substitutions and taking advantage
of trigonometric identities, such as $\sin^2\theta + \cos^2\theta = 1$
or $\tan^2\theta + 1 = \sec^2\theta$.

For example, to evaluate

$$\int \frac{1}{\sqrt{1-x^2}}\,dx$$

make the substitution $x = \sin\theta$. Then $dx = \cos\theta\,d\theta$,
and $\theta = \arcsin x$. The integral becomes:

$$\begin{aligned}
\int \frac{1}{\sqrt{1-\sin^2\theta}}\cos\theta\,d\theta &= \int \frac{1}{\sqrt{\cos^2\theta}}\cos\theta\,d\theta \\
&= \int \frac{1}{\cos\theta}\cos\theta\,d\theta \\
&= \int d\theta = \theta + C
\end{aligned}$$

Therefore:

$$\int \frac{1}{\sqrt{1-x^2}}\,dx = \arcsin x + C$$

The following integral can be solved by making
the substitution $x = \tan\theta$:

$$\int \frac{1}{1+x^2}\,dx = \arctan x + C$$

For another example of this method, see **double
integral.**

INTERCEPT The y-intercept of a curve is the value
of y where it crosses the y-axis, and the x intercept
is the value of x where the curve crosses the x-axis.

For the line $y = mx + b$, the y intercept is b and the x intercept is $-b/m$.

INTERPOLATION Interpolation provides a means of estimating the value of a function for a particular number if you know the value of the function for two other numbers above and below the number in question. For example, $\sin 26° = 0.4384$ and $\sin 27° = 0.4540$. It seems reasonable to suppose that $\sin(26\frac{2}{3}°)$ will be approximately two-thirds of the way between 0.4384 and 0.4540, or 0.4488. This approximation is close to the true value as long as the two numbers you are interpolating between are close to each other. The general formula for interpolation when $a < c < b$ is

$$f(c) = f(a) + \frac{c - a}{b - a}[f(b) - f(a)]$$

INTERSECTION The intersection of two sets is the set of all elements contained in both sets. For example, the intersection of the sets {1,2,3,4,5,6} and {2,4,6,8,10,12} is the set {2,4,6}. William Howard Taft is the only member of the intersection between the set of Presidents of the United States and the set of Chief Justices of the United States. The set of squares is the intersection between the set of rhombuses and the set of rectangles. The intersection of set A and set B is symbolized by $A \cap B$.

INVERSE If $\circ$ represents an operation (such as addition), and I represents the identity element of that operation, then the inverse of a number x is the number y such that $x \circ y = I$. For example, the additive inverse of a number x is $-x$ (also called the *nega-*

tive of x) because $x + (-x) = 0$. The multiplicative inverse of x is $1/x$ (also called the *reciprocal* of x) because $x \cdot \frac{1}{x} = 1$ (assuming $x \neq 0$).

INVERSE FUNCTION An inverse function is a function that does exactly the opposite of the original function. If the function g is the inverse of the function f, and if $y = f(x)$, then $x = g(y)$. For example, the natural logarithm function is the inverse of the exponential function: If $y = e^x$, then $x = \ln y$.

INVERSE MATRIX The inverse of a square matrix **A** is the matrix that, when multiplied by **A**, gives the identity matrix **I**. **A** inverse is written as $\mathbf{A}^{-1}$: $\mathbf{A}\mathbf{A}^{-1} = \mathbf{I}$.

(See **matrix; matrix multiplication; identity matrix**.)

$\mathbf{A}^{-1}$ exists if $\det \mathbf{A} \neq 0$. (See **determinant**.)

The inverse of a 2×2 matrix can be found from the formula:

$$\begin{pmatrix} a & b \\ c & d \end{pmatrix}^{-1} = \begin{pmatrix} \frac{d}{ad-bc} & \frac{-b}{ad-bc} \\ \frac{-c}{ad-bc} & \frac{a}{ad-bc} \end{pmatrix}$$

In general, the element in row i, column j of the inverse matrix can be found from this formula:

element (i,j) in $\mathbf{A}^{-1} =$

$$\frac{a_{ji}^{cofactor}}{\det \mathbf{A}}$$

where

$$a_{ji}^{cofactor} = (-1)^{j+i} \times \det(a_{ji}^{minor})$$

and a_{ij}^{minor} is the matrix formed by crossing out row i and column j in matrix **A**. (See **minor**.) For an application, see **simultaneous equations**.

INVERSE TRIGONOMETRIC FUNCTIONS

The inverse trigonometric functions (figure 62) are six functions (designated with the prefix "arc") that are the inverse functions for the six trigonometric functions:

If $a = \sin b$, then $b = \arcsin a$
If $a = \cos b$, then $b = \arccos a$
If $a = \tan b$, then $b = \arctan a$
If $a = \operatorname{ctn} b$, then $b = \operatorname{arcctn} a$
If $a = \sec b$, then $b = \operatorname{arcsec} a$
If $a = \csc b$, then $b = \operatorname{arccsc} a$

There are many values of b such that $a = \sin b$, for a given a. For example, $\sin(\pi/6) = \sin(2\pi + \pi/6) = \sin(4\pi + \pi/6) = 1/2$. Therefore, it is necessary to specify a range of principal values for each of these functions so that there is only one value of the dependent variable for each value of the independent variable. The name of the function is capitalized to indicate that the principal values are to be taken. The table lists the domain and the range of the principal values for the inverse trigonometric functions.

Function	Inverse Function	Domain	Range (principal values)
$x = \sin y$	$y = \operatorname{Arcsin} x$	$-1 \leq x \leq 1$	$-\pi/2 \leq y \leq \pi/2$
$x = \cos y$	$y = \operatorname{Arccos} x$	$-1 \leq x \leq 1$	$0 \leq y \leq \pi$
$x = \tan y$	$y = \operatorname{Arctan} x$	all real numbers	$-\pi/2 < y < \pi/2$
$x = \operatorname{ctn} y$	$y = \operatorname{Arcctn} x$	all real numbers	$0 < y < \pi$
$x = \sec y$	$y = \operatorname{Arcsec} x$	$|x| \geq 1$	$0 < y < \pi$
$x = \csc y$	$y = \operatorname{Arccsc} x$	$|x| \geq 1$	$-\pi/2 < y < \pi/2$

(Note: the ranges given for arcsecant and arccosecant are chosen to match the ranges of their corresponding reciprocal functions. The range for arcsecant could also be given as $-\pi/2$ to $\pi/2$ so that

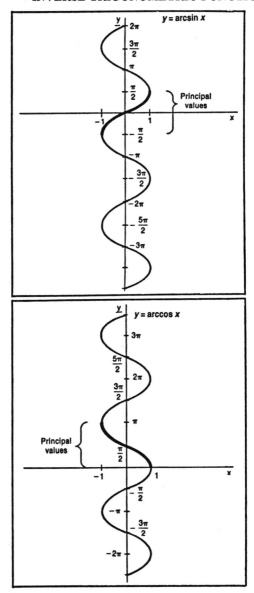

Figure 62 Inverse trigonometric functions

it follows one continuous branch of the curve. The range for arccosecant could likewise be given as 0 to π.)

For example, if you need to walk in a straight line toward a point 4 miles north and 3 miles east, then you need to walk at an angle $\arctan \frac{4}{3} = 53.1$ degrees north of east.

INVERSELY PROPORTIONAL If y and x are related by the equation $y = k/x$, where k is a constant, then y is said to be inversely proportional to x.

IRRATIONAL NUMBER An irrational number is a real number that is not a rational number (i.e., it cannot be expressed as the ratio of two integers). Irrational numbers can be represented by decimal fractions in which the digits go on forever without ever repeating a pattern. Some of the most common irrational numbers are square roots, such as $\sqrt{3} = 1.732050808\ldots$. Also, most values of trigonometric functions are irrational, such as $\sin(10°) = 0.1736481777\ldots$. The special numbers π (pi) and e are also irrational.

To show that $\sqrt{2}$ is not a rational number, we need to show that there are no two integers such that their ratio is $\sqrt{2}$. Suppose that there were two such integers (call them a and b) with no common factors. Then $a^2/b^2 = 2$, so $a^2 = 2b^2$. Therefore a^2 is even (meaning that it is divisible by 2). If a^2 is even, then a itself must be even. This means that a can be expressed as $a = 2c$, where c is also an integer. Then $a^2 = 4c^2 = 2b^2$, or $b^2 = 2c^2$. This means that b^2 is even, and thus b is even. We have reached a contradiction, since we originally assumed that a and b had no common factors. Since we reach a contradiction if we assume that $\sqrt{2}$ is rational, it must be irrational. We can easily find a distance that is $\sqrt{2}$ units long, though. If we draw a right triangle with two sides each one unit long, then the third side will have length $\sqrt{2}$. (See **Pythagorean theorem**.) The radical $\sqrt{2}$ can be approximated by the decimal fraction $1.414213562\ldots$

ISOMETRY An isometry is a way of transforming a figure that does not change the distances between any two points on the figure. For example, a translation or a rotation is an isometry. However, if a figure is transformed by making it twice as big, then the transformation is not an isometry.

ISOSCELES TRIANGLE An isosceles triangle is a triangle with two equal sides.

J

JACOBIAN If $f(x, y)$, $g(x, y)$ are two functions of two variables, then the Jacobian matrix is the matrix of partial derivatives:

$$\begin{pmatrix} \frac{\partial f}{\partial x} & \frac{\partial f}{\partial y} \\ \frac{\partial g}{\partial x} & \frac{\partial g}{\partial y} \end{pmatrix}$$

(The analogous definition also applies to cases with more than two dimensions.) The determinant of this matrix is known as the Jacobian determinant.

JOINT VARIATION If $z = kxy$, where k is a constant, then z is said to vary jointly with x and y.

K

KEPLER Johannes Kepler (1571 to 1630) was a German astronomer who used observational data to express the motion of the planets according to three mathematical laws: (1) planets move along orbits shaped like ellipses, with the sun at one focus; (2) a radius vector connecting the sun to the planet sweeps out equal areas in equal times (this means that a planet travels fastest when closest to the sun); (3) the square of the orbital period is proportional to the cube of the mean distance from the planet to the sun.

KOVALEVSKAYA Sofya Kovalevskaya (1850 to 1891) was a mathematician who worked in Germany and Sweden and made important contributions in differential equations.

L

LAGRANGE Joseph-Louis Lagrange (1736 to 1813) was an Italian-French mathematician who developed ideas in celestial mechanics, calculus of variations, and number theory.

LAPLACE Pierre-Simon Laplace (1749 to 1827) was a French astronomer and mathemetician who investigated the motion of the planets of the solar system.

LATUS RECTUM The latus rectum of a parabola (see figure 63) is the chord through the focus perpendicular to the axis of symmetry. The latus rectum of an ellipse is one of the chords through a focus that is perpendicular to the major axis.

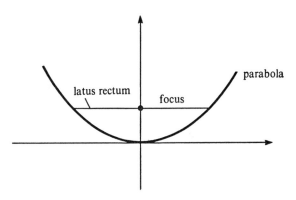

Figure 63

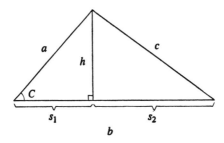

Figure 64 Law of Cosines

LAW OF COSINES The law of cosines (see figure 64) allows us to calculate the third side of a triangle if we know the other two sides and the angle between them:

$$c^2 = a^2 + b^2 - 2ab\cos C$$

In this formula, a, b, and c are the three sides of the triangle, and C is the angle opposite side c.

Calling the altitude of the triangle h, we know from the Pythagorean theorem that $h^2 + s_2^2 = c^2$. Solving for h and s_2 gives:

$$h = a\sin C$$
$$s_2 = b - s_1 = b - a\cos C$$
$$c^2 = a^2\sin^2 C + b^2 - 2ab\cos C + a^2\cos C$$

Using the fact that $\sin^2 C + \cos^2 C = 1$, we obtain

$$c^2 = a^2 + b^2 - 2ab\cos C$$

The final equation is the law of cosines. It is a generalization of the Pythagorean theorem. For $C = 90° = \pi/2$, we have a right triangle with c as the hypotenuse and $\cos C = 0$, so the law of cosines reduces to the Pythagorean theorem.

For example, to calculate the third side of an isosceles triangle with two sides that are 10 units long adjacent to a 100° angle, we use this formula:

$$c^2 = 10^2 + 10^2 - 2 \times 10 \times 10 \times \cos 100°$$
$$c = 15.3$$

(See also **solving triangles**.)

LAW OF LARGE NUMBERS The law of large numbers states that if a random variable is observed many times, the average of these observations will tend toward the expected value (mean) of that random variable. For example, if you roll a die many times and calculate the average value for all of the rolls, you will find that the average value will tend to approach 3.5.

LAW OF SINES The law of sines expresses a relationship involving the sides and angles of a triangle:

$$\frac{a}{\sin A} = \frac{b}{\sin B} = \frac{c}{\sin C}$$

In each case a small letter refers to the length of a side, and a capital letter designates the angle opposite that side. (See figure 65.) The law can be demonstrated by calling h the altitude of the triangle:

$$\frac{h}{b} = \sin A$$
$$\frac{h}{a} = \sin B$$
$$b \sin A = a \sin B$$

A similar demonstration will show that the law works for c and C. (See **solving triangles**.)

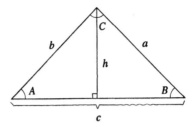

Figure 65 Law of Sines

LAW OF TANGENTS If a, b, and c are the lengths of the sides of a triangle, and A, B, and C are the angles opposite these three sides, respectively, then the law of tangents states that the following relations will be true:

$$\frac{a-b}{a+b} = \frac{\tan[\frac{1}{2}(A-B)]}{\tan[\frac{1}{2}(A+B)]}$$

$$\frac{b-c}{b+c} = \frac{\tan[\frac{1}{2}(B-C)]}{\tan[\frac{1}{2}(B+C)]}$$

$$\frac{c-a}{c+a} = \frac{\tan[\frac{1}{2}(C-A)]}{\tan[\frac{1}{2}(C+A)]}$$

LEAST COMMON DENOMINATOR The least common denominator of two fractions a/b and c/d is the smallest integer that contains both b and d as a factor. For example, the least common denominator of the fractions 3/4 and 5/6 is 12, since 12 is the smallest integer that has both 4 and 6 as a factor.

To add two fractions, turn them both into equivalent fractions whose denominator is the least common denominator. For example, to add $3/4 + 5/6$:

$$\frac{3}{4} = \frac{3}{4} \times \frac{3}{3} = \frac{9}{12}$$

$$\frac{5}{6} = \frac{5}{6} \times \frac{2}{2} = \frac{10}{12}$$

$$\frac{3}{4} + \frac{5}{6} = \frac{9}{12} + \frac{10}{12} = \frac{19}{12}$$

LEAST COMMON MULTIPLE The least common multiple of two natural numbers is the smallest natural number that has both of them as a factor. For example, 6 is the least common multiple of 2 and 3, and 30 is the least common multiple of 10 and 6.

LEAST SQUARES ESTIMATOR See **regression; multiple regression.**

LEIBNIZ Gottfried Wilhelm Leibniz (1646 to 1716) was a German mathematician, philosopher, and political advisor, who was one of the developers of calculus (independently of his rival Newton).

LEMMA A lemma is a theorem that is proved mainly as an aid in proving another theorem.

LEVEL OF SIGNIFICANCE The level of significance for a hypothesis-testing procedure is the probability of committing a type 1 error. (See **hypothesis testing.**)

L'HOSPITAL'S RULE L'Hospital's rule tells how to find the limit of the ratio of two functions in cases where that ratio approaches $0/0$ or ∞/∞. Let y represent the ratio between two functions, $f(x)$ and $g(x)$:

$$y = \frac{f(x)}{g(x)}$$

Then l'Hospital's rule states that

$$\lim_{x \to a} y = \frac{\lim_{x \to a} f'(x)}{\lim_{x \to a} g'(x)}$$

where $f'(x)$ and $g'(x)$ represent the derivatives of these functions with respect to x.

For example, suppose that

$$y = \frac{2x^2 + 18x - 44}{2x - 4}$$

and we need to find $\lim_{x \to 2} y$. We cannot find this limit directly, because inserting the value $x = 2$ in the expression for y gives the expression $0/0$. However, by setting $f(x) = 2x^2 + 18x - 44$, we can find $f'(x) = 4x + 18$, $\lim_{x \to 2} f'(x) = 26$, $g(x) = 2x - 4$, $g'(x) = 2$.

Therefore,

$$\lim y_{x \to 2} = \frac{26}{2} = 13$$

For another example, suppose that

$$y = \frac{Pr(1 + r)^n}{(1 + r)^n - 1}$$

and assume that n and P are constant. To find $\lim_{r \to 0} y$, we must use l'Hospital's rule. We let

$$f(r) = Pr(1+r)^n; \quad f'(r) = Prn(1+r)^{n-1} + P(1+r)^n$$

$$\lim_{r \to 0} f'(r) = P$$

$$g(r) = (1+r)^n - 1; \quad g'(r) = n(1+r)^{n-1}; \quad \lim_{r \to 0} g'(r) = n$$

Therefore

$$\lim_{r \to 0} y = \frac{P}{n}$$

This formula represents the monthly payment for a home mortgage, where r is the monthly interest rate, n is the number of months to repay the loan, and P is the principal amount (the amount that is borrowed). The result says that, if the interest rate is zero, the monthly payment is simply equal to the principal amount divided by the number of months.

LIKE TERMS Two terms are like terms if all parts of both terms except for the numerical coefficients are the same. For example, the terms $3a^2b^3c^4$ and $-6.5a^2b^3c^4$ are like terms. If two like terms are added, they can be combined into one term. For example, the sum of the two terms above is $-3.5a^2b^3c^4$.

LIMIT The limit of a function is the value that the dependent variable approaches as the independent variable approaches some fixed value. The expression "The limit of $f(x)$ as x approaches a" is written as

$$\lim_{x \to a} f(x)$$

For example:

$$\lim_{x \to 2} x^2 = 4, \ \lim_{x \to \pi/2} \sin x = 1, \ \lim_{x \to 1} x^2 + 3x + 1 = 5$$

In each of these cases the limit is not very interesting, because we can easily find $f(2)$, $f(\pi/2)$, or $f(1)$. However, there are cases where $\lim_{x \to a} f(x)$ exists, but $f(a)$ does not. For example:

$$f(x) = \frac{(x-1)(x+2)}{x-1}$$

is undefined if $x = 1$. However, the closer that x comes to 1, the closer $f(x)$ approaches 3. For example, $f(1.0001) = 3.0001$. All of calculus is based on this type of limit. (See **derivative**.)

The formal definition of limit is: The limit of $f(x)$ as x approaches a exists and is equal to B if, for any positive number ϵ (no matter how small), there exists a positive number δ such that, if $0 < |x-a| < \delta$, then $|f(x) - B| < \epsilon$.

LINE A line is a straight set of points that extends off to infinity in two directions. The term "line" is one of the basic undefined terms in Euclidian geometry, so it is not possible to give a rigorous definition of line. You will have to use your intuition as to what it means for a line to be straight. According to a postulate, any two distinct points determine one and only one line. A line has infinite length, but zero width and zero thickness. (See also **line segment**.)

LINE INTEGRAL Let **E** be a three-dimensional vector field, and let Δ **L** be a small vector representing a portion of a path L in three-dimensional space. Take the dot product $\mathbf{E} \cdot \Delta \mathbf{L}$, and then add up all of these products for all elements of the path; now, take the limit as the length of each path segment goes to zero, and you have the line integral of the field **E** along the path L:

$$\int_{path} \mathbf{E} \cdot \mathbf{dL}$$

In order to evaluate the integral, the path needs to be expressed in terms of some parameter, and then the limits of integration are given in terms of this parameter. The examples below are chosen so that

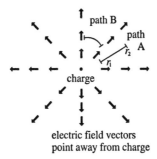

electric field vectors
point away from charge

Figure 66

the paths are relatively simple.

Let $\mathbf{E}$ be a vector field with magnitude given by:

$$\|\mathbf{E}\| = \frac{q}{4\pi\epsilon_0 r^2}$$

whose direction always point away from the origin.
(This is the electric field created by a point electric
charge with charge q located at the origin.)

Consider a line integral along a path radially outward from the charge, starting at distance r_1 and
ending at distance r_2. (See path A in figure 66.) In
this case the field vector $\mathbf{E}$ points in the same direction as the path vector $\mathbf{dL}$, so the dot product
between them will simply be the product of their
magnitudes:

$$\mathbf{E} \cdot \mathbf{dL} = \|\mathbf{E}\| \times \|\mathbf{dL}\| = \frac{q}{4\pi\epsilon_0 r^2} dr$$

(We can rename dL as dr because this path is only in the direction of increasing r.) The line integral
becomes:

$$\int_{r_1}^{r_2} \frac{q}{4\pi\epsilon_0 r^2} dr$$

$$= \frac{q}{4\pi\epsilon_0} \int_{r_1}^{r_2} r^{-2} dr$$

$$= \frac{q}{4\pi\epsilon_0} (-r^{-1})|_{r_1}^{r_2}$$

$$= \frac{q}{4\pi\epsilon_0} (1/r_1 - 1/r_2)$$

Now, consider an example of a line integral along a circle that is centered at the origin. (See path B in figure 66.) In this case, the field vector $\mathbf{E}$ is everywhere perpendicular to the path vector $\mathbf{dS}$, so the dot product is everywhere 0. Therefore, the line integral is zero.

Any arbitrary path can be broken into tiny segments, some of which are arcs of circles centered on the origin, and others which travel radially outward or inward. The circular parts will contribute 0 to the total line integral, and the total contribution of the radial parts will depend only on the distances r_1 and r_2. In particular, if you take the line integral of the electric field along any closed path (i.e, a path that ends up at the same place it started), then $r_1 = r_2$ and the value of the integral will be zero. There are important implications when a vector field has this special property. (See **potentional function; Stokes's theorem; Maxwell's equations.**)

For another example, let the vector field $\mathbf{B}$ be defined by:

$$\mathbf{B}(x, y) = \left[\frac{-y\mu_0 I}{x^2 + y^2}, \frac{x\mu_0 I}{x^2 + y^2} \right]$$

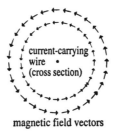

magnetic field vectors

Figure 67

where μ_0 and I are constants. The field vector at any point will be perpendicular to the vector connecting the origin to that point. (See figure 67.)

(This field represents the magnetic field generated by a current I flowing through a long, straight wire along the z axis. The field does not change as z changes, so we have not explicitly included the z coordinate.)

The magnitude of the field is $\mu_0 I/r$. Written in polar coordinates, the field is:

$$\mathbf{B}(r,\theta) = \frac{\mu_0 I}{r}\hat{\theta}$$

where $\hat{\theta}$ is a unit vector pointing in the direction of the field.

Now, take the line integral of the magnetic field along a circular path centered on the wire. In each case the field vector $\mathbf{B}$ points in the same direction as the path vector $\mathbf{dL}$, so the dot product is simply the product of their magnitudes:

$$\mathbf{B} \cdot \mathbf{dL} = \|\mathbf{B}\| \cdot \|\mathbf{dL}\|$$

We can write $\|\mathbf{dL}\|$ as $r d\theta$, and the line integral around the entire circle can be written:

$$\int_0^{2\pi} \frac{\mu_0 I}{2\pi r} r\,d\theta$$

$$= \frac{\mu_0 I}{2\pi} \int_0^{2\pi} d\theta$$

$$= \mu_0 I$$

If we take the line integral along a path that goes radially outward from the wire, then the field vector **B** will be everywhere perpendicular to the path vector **dL**, so the dot product between them will be zero.

LINE SEGMENT A line segment is like a piece of a line. It consists of two end points and all of the points on the straight line between those two points.

LINEAR COMBINATION A linear combination of two vectors **x** and **y** is a vector of the form $a\mathbf{x} + b\mathbf{y}$, where a and b are scalars. (See also **linearly independent.**)

LINEAR EQUATION A linear equation with unknown x is an equation that can be written in the form $ax + b = 0$. For example, $2x - 10 = 2$ can be written as $2x - 12 = 0$, so this is a linear equation with the solution $x = 6$. (See **simultaneous equations.**)

LINEAR FACTOR A linear factor is a factor that includes only the first power of an unknown. For example, in the expression $y = (x - 2)(x^2 + 3x + 4)$, the factor $(x - 2)$ is a linear factor, but the factor $(x^2 + 3x + 4)$ is a quadratic factor.

LINEAR PROGRAMMING A linear programming problem is a problem for which you need to choose the optimal set of values for some variables subject to some constraints. The goal is to maximize or minimize a function called the *objective function*. In a linear programming problem, the objective function and the constraints must all be linear functions; that is, they cannot involve variables raised to any power (other than 1), and they cannot involve two variables being multiplied together.

Some examples of problems to which linear programming can be applied include finding the least-cost method for producing a given product, or finding the revenue-maximizing product mix for a production facility with several capacity limitations.

Here is an example of a linear programming problem:

Maximize $6x + 8y$ subject to:

$$y \leq 10$$
$$x + y \leq 15$$
$$2x + y \leq 25$$
$$x \geq 0$$
$$y \geq 0$$

This problem has two choice variables: x and y. The objective function is $6x + 8y$, and there are three constraints (not counting the two nonnegativity constraints $x \geq 0$ and $y \geq 0$).

It is customary to rewrite the constraints so that they contain equals signs instead of inequality signs. In order to do this some new variables, called *slack variables*, are added. One slack variable is added for each constraint. Here is how the problem given above

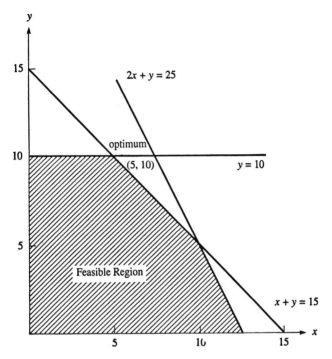

Figure 68 Linear programming

looks when three slack variables (s_1, s_2, and s_3) are included.

Maximize $6x + 8y$ subject to:

$$y + s_1 = 10$$
$$x + y + s_2 = 15$$
$$2x + y + s_3 = 25$$
$$x \geq 0, \ y \geq 0, \ s_l \geq 0, \ s_2 \geq 0, \ s_3 \geq 0$$

Each slack variable represents the excess capacity associated with the corresponding constraint.

The feasible region consists of all points that satisfy the constraints. (See figure 68.) A theorem of linear programming states that the optimal solution will lie at one of the corner points of the feasible region. In this case the optimal solution is at the point $x = 5, y = 10$.

A linear programming problem with two choice variables can be solved by drawing a graph of the feasible region, as was done above. If there are more than two variables, however, it is not possible to draw a graph, and the problem must then be solved by an algebraic procedure, such as the **simplex method**.

LINEARLY INDEPENDENT A set of vectors **a**, **b**, and **c** is linearly independent if it is impossible to find three scalars m, n, and p (not all zero) such that $m\mathbf{a}+n\mathbf{b}+p\mathbf{c} = \mathbf{0}$. Two vectors clearly are not linearly independent if they are multiples of each other; for example, if $\mathbf{a} = (2,3,4)$ and $\mathbf{b} = (20,30,40)$, then $10\mathbf{a} - \mathbf{b} = \mathbf{0}$. With a set of three vectors, it can be more complicated to tell. For example, let $\mathbf{a} = (2,3,4)$, $\mathbf{c} = (5,6,7)$, and $\mathbf{d} = (19,24,29)$. No two of these vectors are multiples of each other. Still, they are not linearly independent, because $2\mathbf{a} + 3\mathbf{b} - \mathbf{d} = \mathbf{0}$. If the vectors are arranged as the columns of a square matrix, then they are linearly independent if and only if the determinant of that matrix is not zero. (See **determinant; rank**.) In this case, the determinant

$$\begin{vmatrix} 2 & 5 & 19 \\ 3 & 6 & 24 \\ 4 & 7 & 29 \end{vmatrix}$$

is zero.

LITERAL A literal number is a number expressed as a numeral, not as a variable. For example, in the equation $x = 2.4y$, 2.4 is a literal number.

LN See **natural logarithm**.

LOBACHEVSKY Nikolay Lobachevsky (1792 to 1856) was a Russian mathematician who developed a version of non-Euclidian geometry.

LOCUS The term "locus" is a technical way of saying "set of points." For example, a circle can be defined as being "the locus of points in a plane that are a fixed distance from a given point." The plural of "locus" is "loci."

LOG The function $y = \log x$ is an abbreviation for the logarithm function to the base 10. (See **logarithm**.)

LOGARITHM A logarithm is an inverse of an exponential. The equation $x = a^y$ can be written as $y = \log_a x$, which means "y is the logarithm to the base a of x." Any positive number (except 1) can be used as the base for a logarithm function. The two most useful bases are 10 and e. Logarithms to the base 10 are called *common logarithms*. They are very convenient to use, since we use a base 10 number system. If no base is specified in the expression $\log x$, then base 10 is usually meant: $\log x = \log_{10} x$. Here are some examples:

$$
\begin{aligned}
\log 1 &= 0 \quad (\text{because } 10^0 = 1) \\
\log 10 &= 1 \\
\log 100 &= 2 \\
\log 1000 &= 3
\end{aligned}
$$

Except in a few simple cases, logarithms will be irrational numbers. Table 1 lists some values of the common logarithm function.

Logarithms to any base satisfy these properties:

$$
\begin{aligned}
\log(xy) &= \log x + \log y \\
\log(y/x) &= \log y - \log x \\
\log(x^n) &= n \log x
\end{aligned}
$$

These properties follow directly from the properties of exponents.

Logarithms are convenient if we have to measure very large quantities and very small quantities at the same time. For example, the stellar magnitude system for measuring the brightness of stars is based on a logarithmic scale.

Logarithms have also been very helpful as calculation aids, because a multiplication problem can be turned into an addition problem by taking the logarithms. (See **slide rule**.) However, this use has become less important as pocket calculators have become widely available.

Logarithms to the base e are important in calculus. (See **natural logarithm**.)

LOGIC Logic is the study of sound reasoning. The study of logic focuses on the study of arguments. An argument is a sequence of sentences (called *premises*), that lead to a resulting sentence (called the *conclusion*). An argument is a valid argument if the conclusion does follow from the premises. In other words, if an argument is valid and all its premises are true, then the conclusion must be true.

Here is an example of a valid argument:

Premise: If a shape is a square, then it is both a rectangle and a rhombus.
Premise: Central Park is not a rhombus.
Conclusion: Therefore, Central Park is not a square.

Here is another example of an argument:

Premise: If a shape is either a rhombus or a rectangle, then it is a square.
Premise: Central Park is a rectangle.
Conclusion: Therefore, Central Park is a square.

This is a valid argument, since the conclusion follows from the premises. However, one of the premises (the first one) is false. If any of the premises of an argument is false, then the argument is called an unsound argument.

Logic can be used to determine whether an argument is valid; however, logic alone cannot determine whether the premises are true or false. Once an argument has been shown to be valid, then all other arguments of the same general form will also be valid, even if their premises are different.

Arguments are composed of sentences. Sentences are said to have the truth value T (corresponding to what we normally think of as "true") or the truth value F (corresponding to "false"). In studying the general logical properties of sentences, it is customary to represent a sentence by a lower-case letter, such as p, q, or r, called a sentence variable or a Boolean variable. Sentences either can be simple sentences or can consist of simple sentences joined by connectives and called compound sentences. For example, "Spot is a dog" is a simple sentence. "Spot is a dog and Spot likes to bury bones" is a compound sentence. The connectives used in logic include AND, OR, and NOT. To learn how these are used, see **Boolean algebra.**

M

MACLAURIN Colin Maclaurin (1698 to 1746) was a Scottish mathematician who extended the field of calculus. (See **Maclaurin series**.)

MACLAURIN SERIES The Maclaurin series is a special case of the Taylor series for $f(x + h)$, when $x = 0$. (See **Taylor series**.)

MAGNITUDE The magnitude of a vector **a** is its length. It is symbolized by two pairs of vertical lines, and it can be found by taking the square root of the dot product of the vector with itself:

$$\|\mathbf{a}\| = \sqrt{\mathbf{a} \cdot \mathbf{a}}$$

For example, the magnitude of the vector (3, 4) is

$$\sqrt{(3,4) \cdot (3,4)} = \sqrt{9 + 16} = \sqrt{25} = 5$$

MAJOR ARC A major arc of a circle is an arc with a measure greater than 180°. (See **arc**.)

MAJOR AXIS The major axis of an ellipse is the line segment joining two points on the ellipse that passes through the two foci. It is the longest possible distance across the ellipse. (See **ellipse**.)

MAJOR PREMISE The major premise is the sentence in a syllogism that asserts a general relationship between classes of objects. (See **syllogism**.)

MANDELBROT SET The Mandelbrot set, discovered by Benoit Mandelbrot, is a famous fractal, i.e.,

$X = -2.00$ to 1.25 $Y = -1.50$ to 1.50 $X = -0.30$ to 0.00 $Y = -0.85$ to 1.10

Figure 69 Mandelbrot set

a shape containing an infinite amount of fine detail. It is the set of values of c for which the series $z_{n+1} = z_n^2 + c$ converges, where z and c are complex numbers and z is initially $(0,0)$. (See **complex number**.)

Figure 69 shows the whole set and an enlargement of a small area. On the plot, x and y are the real and imaginary parts of c. The Mandelbrot set is the black bulbous object in the middle; elsewhere, the stripes indicate the number of iterations needed to make $|z|$ exceed 2.

MANTISSA The mantissa is the part of a common logarithm to the right of the decimal point. For example, in the expression $\log 115 = 2.0607$, the quantity 0.0607 is the mantissa. For contrast, see **characteristic**.

MAPPING A mapping is a rule that, to each member of one set, assigns a unique member of another set.

MATHEMATICAL INDUCTION Mathematical induction is a method for proving that a proposition is true for all whole numbers. First, show that the proposition is true for a few small numbers, such as 1, 2, and 3. Then show that, if the proposition is true for an arbitrary number j, then it must be true for the next number: $j + 1$. Once you have done these two steps, the proposition has been proved, since, if it is true for 1, then it must also be true for 2, which means it must be true for 3, which means it must be true for 4, and so on.

For example, we can prove that

$$\sum_{i=1}^{n} i = 1 + 2 + 3 + 4 + \cdots + n = \frac{n(n+1)}{2}$$

is true for all natural numbers n.

(See **arithmetic series; summation notation**.) The proposition is true for $n = 1, n = 2$, and $n = 3$:

$$\sum_{i=1}^{1} i = 1 = \frac{1(1+1)}{2}$$

$$\sum_{i=1}^{2} i = 1 + 2 = 3 = \frac{2(2+1)}{2}$$

$$\sum_{i=1}^{3} i = 1 + 2 + 3 = 6 = \frac{3(3+1)}{2}$$

Now assume that this formula is true for any arbitrary natural number j. Then:

$$\sum_{i=1}^{j+1} i = \sum_{i=1}^{j} i + (j+1)$$

$$= \frac{j(j+1)}{2} + (j+1)$$

$$= \frac{j^2 + j + 2j + 2}{2} = \frac{(j+2)(j+1)}{2}$$

Therefore the formula works for $j + 1$ if it works for j, so it must be true for all j.

MATHEMATICS Mathematics is the orderly study of the structures and patterns of abstract entities. Normally the objects that mathematicians talk about correspond to objects about which we have an intuitive understanding. For example, we have an intuitive notion of what a number is, what a line in three-dimensional space is, and what the concept of probability is.

Applied mathematics is the field in which mathematical concepts are applied to practical problems. For example, the lines and points that pure mathematics deals with are abstractions that we can't see or touch. However, these abstract ideas correspond closely to the concrete objects that we think of as lines or points. Mathematics was originally developed for its applied value. The ancient Egyptians and Babylonians developed numerous properties of numbers and geometric figures that they used to solve practical problems.

The formal procedure of mathematics is this: Start with some concepts that will be left undefined, such as "number" or "line." Then make some postulates that will be assumed to be true, such as "Every natural number has a successor." Next make definitions using undefined terms and previously defined terms, such as "A circle is the set of all points in a plane that are a fixed distance from a given point." Then use the postulates to prove theorems, such as the Pythagorean theorem. Once a theorem has been proved, it can then be used in the proof of other theorems.

MATRIX A matrix is a table of numbers arranged in rows and columns. The plural of "matrix" is "matrices." The size of a matrix is characterized by two numbers: the number of rows and the number of columns. Matrix **A** is a 2×2 matrix, matrix **B** is 3×2, matrix **C** is 3×3, and matrix **D** is 2×3:

$$\mathbf{A} = \begin{pmatrix} 1 & 2 \\ 3 & 4 \end{pmatrix}$$

$$\mathbf{B} = \begin{pmatrix} 0 & 6 \\ 10 & 5 \\ 4 & 2 \end{pmatrix}$$

$$\mathbf{C} = \begin{pmatrix} 1 & 0 & 1 \\ 0 & 1 & 0 \\ 1 & 0 & 1 \end{pmatrix}$$

$$\mathbf{D} = \begin{pmatrix} 100 & 15 & 25 \\ 36 & 10 & 15 \end{pmatrix}$$

(The number of rows is always listed first.) A baseball box score is an example of a 9×4 matrix.

	ab	r	h	rbi
shortstop	5	3	3	0
first baseman	4	1	2	1
right fielder	4	0	1	2
center fielder	4	0	1	0
left fielder	4	0	0	0
catcher	4	1	1	1
third baseman	4	0	1	0
pitcher	3	0	0	0
second baseman	3	0	1	0

The transpose of a matrix **A** (written as $\mathbf{A}^{tr}$ or $\mathbf{A}'$) is formed by turning all the rows into columns and all the columns into rows. For example, the transpose of $\begin{pmatrix} 11 & 12 \\ 21 & 22 \\ 31 & 32 \end{pmatrix}$ is the matrix $\begin{pmatrix} 11 & 21 & 31 \\ 12 & 22 & 32 \end{pmatrix}$

Matrices can be multiplied by the rules of matrix multiplication. If **A** is an $m \times n$ matrix, and **B** is an $n \times p$ matrix, then the product **AB** will be an $m \times p$ matrix. The product **AB** can be found only if the number of columns in matrix **A** is equal to the number of rows in matrix **B**. (See **matrix multiplication**.) A square matrix is a matrix in which the numbers of rows and columns are equal. One important square matrix is the matrix with ones all along the diagonal from the upper left-hand corner to the lower right-hand corner, and zeros everywhere else. This type of matrix is called an identity matrix, written as **I**. For example, here is a 3×3 identity matrix:

$$\begin{pmatrix} 1 & 0 & 0 \\ 0 & 1 & 0 \\ 0 & 0 & 1 \end{pmatrix}$$

An identity matrix has the important property that, whenever it multiplies another matrix, it leaves the other matrix unchanged: $\mathbf{IA} = \mathbf{A}$.

For many square matrices there exists a special matrix called the inverse matrix (written as $\mathbf{A}^{-1}$), which satisfies the special property that $\mathbf{A}^{-1}\mathbf{A} = \mathbf{I}$. (See **inverse matrix.**)

The determinant of a square matrix (written as det $\mathbf{A}$) is a number that characterizes some important properties of the matrix. If det $\mathbf{A} = 0$, then $\mathbf{A}$ does not have an inverse matrix.

The trace of a square matrix is the sum of the diagonal elements of the matrix. For example, the trace of a 3×3 identity matrix is 3.

The use of matrix multiplication makes it easier to express linear simultaneous equation systems. The system of equations can be written as $\mathbf{Ax} = \mathbf{b}$, where $\mathbf{A}$ is an $m \times m$ matrix of coefficients, $\mathbf{x}$ is an $m \times 1$ matrix of unknowns, and $\mathbf{b}$ is an $m \times 1$ matrix of known constants. If you know $\mathbf{A}^{-1}$, you can find the solution for $\mathbf{x}$:

$$\begin{aligned} \mathbf{Ax} &= \mathbf{b} \\ \mathbf{A}^{-1}\mathbf{Ax} &= \mathbf{A}^{-1}\mathbf{b} \\ \mathbf{Ix} &= \mathbf{A}^{-1}\mathbf{b} \\ \mathbf{x} &= \mathbf{A}^{-1}\mathbf{b} \end{aligned}$$

MATRIX MULTIPLICATION The formal definition of matrix multiplication is as follows:

$$\begin{pmatrix} a_{11} & a_{12} & a_{13} & \cdots & a_{1n} \\ a_{21} & a_{22} & a_{23} & \cdots & a_{2n} \\ & & \vdots & & \\ a_{m1} & a_{m2} & a_{m3} & \cdots & a_{mn} \end{pmatrix}$$

$$\times \begin{pmatrix} b_{11} & b_{12} & b_{13} & \ldots & b_{1p} \\ b_{21} & b_{22} & b_{23} & \ldots & b_{2p} \\ & & \vdots & & \\ b_{n1} & b_{n2} & b_{n3} & \ldots & b_{np} \end{pmatrix}$$

$$= \begin{pmatrix} \sum_{i=1}^{n} a_{1i}b_{i1} & \sum_{i=1}^{n} a_{1i}b_{i2} & \ldots & \sum_{i=1}^{n} a_{1i}b_{ip} \\ \sum_{i=1}^{n} a_{2i}b_{i1} & \sum_{i=1}^{n} a_{2i}b_{i2} & \ldots & \sum_{i=1}^{n} a_{2i}b_{ip} \\ & \vdots & & \\ \sum_{i=1}^{n} a_{mi}b_{i1} & \sum_{i=1}^{n} a_{mi}b_{i2} & \ldots & \sum_{i=1}^{n} a_{mi}b_{ip} \end{pmatrix}$$

Two matrices can be multiplied only if the number of columns in the left-hand matrix is equal to the number of rows in the righthand matrix. If $\mathbf{A}$ is an $m \times n$ matrix (m rows and n columns) and $\mathbf{B}$ is an $n \times p$ matrix, then the product matrix $\mathbf{AB}$ exists and has m rows and p columns. Immediately we can see that matrix multiplication is not commutative, since it makes a difference which matrix is on the left and which is on the right.

The formula for matrix multiplication looks very complicated, but we can make more sense of it by using the dot product of two vectors. The dot product of two vectors is formed by multiplying together each pair of corresponding components and then adding the results of all these products. (See **dot product**.)

A matrix can be thought of either as a vertical stack of row vectors:

$$\begin{pmatrix} a_{11} & \ldots & a_{1n} \\ & \vdots & \\ a_{m1} & \ldots & a_{mn} \end{pmatrix} = \begin{pmatrix} \mathbf{a}_1 \\ \vdots \\ \mathbf{a}_m \end{pmatrix}$$

$$\mathbf{a}_1 = (a_{11}, a_{12}, \ldots, a_{1n})$$

$$\vdots$$

$$\mathbf{a}_m = (a_{m1}, a_{m2}, \ldots, a_{mn})$$

or as a horizontal stack of column vectors:

$$\begin{pmatrix} b_{11} & \cdots & b_{1p} \\ & \vdots & \\ b_{n1} & \cdots & b_{np} \end{pmatrix} = (\ \mathbf{b}_1, \quad \mathbf{b}_2, \quad \ldots, \quad \mathbf{b}_p\ )$$

$$\mathbf{b}_1 = \begin{pmatrix} b_{11} \\ \vdots \\ b_{n1} \end{pmatrix} \ldots \mathbf{b}_p = \begin{pmatrix} b_{1p} \\ \vdots \\ b_{np} \end{pmatrix}$$

For our purposes it is best to think of the left-hand matrix ($\mathbf{A}$) as a collection of row vectors, and the right-hand matrix ($\mathbf{B}$) as a collection of column vectors. Then each element in the matrix product $\mathbf{AB}$ can be found as a dot product of one row of $\mathbf{A}$ with one column of $\mathbf{B}$:

$$\mathbf{AB} =$$

$$\begin{pmatrix} \mathbf{a}_1 \cdot \mathbf{b}_1 & \mathbf{a}_1 \cdot \mathbf{b}_2 & \mathbf{a}_1 \cdot \mathbf{b}_3 & \cdots & \mathbf{a}_1 \cdot \mathbf{b}_p \\ \mathbf{a}_2 \cdot \mathbf{b}_1 & \mathbf{a}_2 \cdot \mathbf{b}_2 & \mathbf{a}_2 \cdot \mathbf{b}_3 & \cdots & \mathbf{a}_2 \cdot \mathbf{b}_p \\ & & \vdots & & \\ \mathbf{a}_m \cdot \mathbf{b}_1 & \mathbf{a}_m \cdot \mathbf{b}_2 & \mathbf{a}_m \cdot \mathbf{b}_3 & \cdots & \mathbf{a}_m \cdot \mathbf{b}_p \end{pmatrix}$$

The element in position $(1, 1)$ of the product matrix is the dot product of the first row of $\mathbf{A}$ with the first column of $\mathbf{B}$. In general, the element in position (i, j) is formed by the dot product of row i in $\mathbf{A}$ and column j in **B**. Examples of matrix multiplication are:

$$\begin{pmatrix} a & b \\ c & d \end{pmatrix} \begin{pmatrix} e & f \\ g & h \end{pmatrix} = \begin{pmatrix} ae+bg & af+bh \\ ce+dg & cf+dh \end{pmatrix}$$

$$\begin{pmatrix} 11 & 12 & 13 \\ 21 & 22 & 23 \\ 31 & 32 & 33 \end{pmatrix} \begin{pmatrix} 1 & 0 \\ 100 & 1 \\ 10000 & 2 \end{pmatrix} = \begin{pmatrix} 131211 & 38 \\ 232221 & 68 \\ 333231 & 98 \end{pmatrix}$$

Matrix multiplication is a very valuable tool, making it much easier to write systems of linear simultaneous equations. The three-equation system

$$a_1x + b_1y + c_1z = d_1$$

$$a_2x + b_2y + c_2z = d_2$$

$$a_3x + b_3y + c_3z = d_3$$

can be rewritten using matrix multiplication as

$$\begin{pmatrix} a_1 & b_1 & c_1 \\ a_2 & b_2 & c_2 \\ a_3 & b_3 & c_3 \end{pmatrix} \begin{pmatrix} x \\ y \\ z \end{pmatrix} = \begin{pmatrix} d_1 \\ d_2 \\ d_3 \end{pmatrix}$$

(See **simultaneous equations**.)

MAXIMA The maxima are the points where the value of a function is greater than it is at the surrounding points. (See **extremum**.)

MAXIMUM LIKELIHOOD ESTIMATOR A maximum likelihood estimator has this property: if the true value of the unknown parameter is the same as the value of the maximum likelihood estimator, then the probability of obtaining the sample that was actually observed is maximized. (See **statistical inference**.)

MAXWELL'S EQUATIONS Maxwell's four equations govern electric and magnetic fields. They were put together by James Clerk Maxwell in the 1870s on the basis of experimental data. These equations can be used to establish the wave nature of light.

First, here are the equations for free space in integral form, assuming there is no change in current over time.

Let **E** be an electric field (a three-dimensional vector field.) These two equations apply:

$$(1) \qquad \int_{closed\ path} \mathbf{E} \cdot \mathbf{dL} = 0$$

(That is, the line integral of the electric field over any closed path is zero.)

$$(2) \qquad \iint_{closed\ surface} \mathbf{E} \cdot \mathbf{dS} = \frac{q_{inside}}{\epsilon_0}$$

(That is, the surface integral of the electric field around any closed surface is equal to q, the total charge inside the surface, divided by a constant known as ϵ_0.)

Let **B** be a magnetic field (also a three-dimensional vector field). Then the line integral around a closed path depends on the current flowing through the interior of the path:

$$(3) \qquad \int_{path\ L} \mathbf{B} \cdot \mathbf{dL} = \mu_0 I_{inside}$$

where I stands for the amount of electric current, and μ_0 is a constant. The surface integral of **B** over a closed surface is zero:

$$(4) \qquad \iint_{closed\ surface} \mathbf{B} \cdot \mathbf{dS} = 0$$

The four equations given above can also be written in alternate forms. Use **Stokes's theorem** to rewrite the two equations involving line integrals. Equation (1) becomes:

$$(5) \qquad \nabla \times \mathbf{E} = \mathbf{0}$$

In words: the curl of the electric field is always zero. Equation (3) becomes:

$$(6) \qquad \nabla \times \mathbf{B} = \mu_0 \mathbf{J}$$

where $\mathbf{J}$, called the current density is defined by this integral:

$$\iint_{surface\ S} \mathbf{J} \cdot \mathbf{dS} = I_{inside}$$

By using the **divergence theorem**, the two equations involving surface integrals can be rewritten. The left-hand side of equation (2) is changed from

$$\iint_{surface\ S} \mathbf{E} \cdot \mathbf{dS}$$

into

$$\iiint_{interior\ of\ S} (\nabla \cdot \mathbf{E}) dV$$

The right-hand side of equation (2) is changed by defining ρ, called the charge density, so that the triple integral of ρ over any volume is equal to the total charge inside that volume:

$$q_{inside\ S} = \iiint_{interior\ of\ S} \rho dV$$

We then have the equation

$$\iiint_{interior\ of\ S} (\nabla \cdot \mathbf{E})\, dV = \iiint_{interior\ of\ S} \left(\frac{\rho}{\epsilon_0}\right) dV$$

Since this equation must hold true for any arbitrary surface S, we can write the equation in this form:

$$(7) \qquad \nabla \cdot \mathbf{E} = \frac{\rho}{\epsilon_0}$$

In words: the divergence of the electric field is proportional to the charge density.

Equation (4) becomes:

$$(8) \qquad \nabla \cdot \mathbf{B} = 0$$

MEAN The mean of a random variable is the same as its **expectation**. The mean of a group of numbers is the same as its **arithmetic mean,** or average.

MEAN VALUE THEOREM If the derivative of a function f is defined everywhere between two points, $(a, f(a))$ and $(b, f(b))$, then the mean value theorem states that there will be at least one value of x between a and b such that the value of the derivative is equal to the slope of the line between $(a, f(a))$ and $(b, f(b))$. This means that there is at least one point in the interval where the tangent line to the curve is parallel to the secant line that passes through the curve at the two endpoints of the interval.

MEDIAN (1) The median of a group of n numbers is the number such that just as many numbers are greater than it as are less than it. For example, the median of the set of numbers $\{1, 2, 3\}$ is 2; the median of $\{1, 1, 1, 2, 10, 15, 16, 20, 100, 105, 110\}$ is 15. In order to determine the median, the list should be placed in numerical order. If there is an odd number of items in the list, then the median is the element in the exact middle. If there is an even number, then the median is the average of the two numbers closest to the middle.

(2) A median of a triangle is a line segment connecting one vertex to the midpoint of the opposite side. (See **triangle**.)

METALANGUAGE A metalanguage is a language that is used to describe other languages.

MIDPOINT Point B is the midpoint of the segment AC if it is between A and C and if $AB = BC$ (that is, the distance from B to A is the same as the distance from B to C).

MINIMA The minima are the points where the value of a function is less than it is at the surrounding points. (See **extremum**.)

MINOR The minor of an element in a matrix is the determinant of the matrix formed by crossing out the row and column containing that element. For example, the minor of the element d in

$$\begin{pmatrix} a & b & c \\ d & e & f \\ g & h & i \end{pmatrix}$$

is the determinant

$$\begin{vmatrix} b & c \\ h & i \end{vmatrix} = bi - ch$$

(See **determinant; inverse matrix**.)

MINOR ARC A minor arc of a circle is an arc with a measure less than $180°$. (See **arc**.)

MINOR AXIS The minor axis of an ellipse is the line segment that passes through the center of the ellipse that is perpendicular to the major axis.

MINOR PREMISE The minor premise is the sentence in a syllogism that asserts a property about a specific case. (See **syllogism**.)

MINUTE A minute is a unit of measure for small angles equal to 1/60 of a degree.

MODE The mode of a group of numbers is the number that occurs most frequently in that group. For example, the mode of the set $\{0, 1, 1, 2, 2, 2, 3, 3, 3, 3, 5, 5, 6, 6, 6\}$ is 3, since 3 occurs four times.

MODULUS The modulus of a complex number is the same as its absolute value.

MONOMIAL A monomial is an algebraic expression that does not involve any additions or subtractions. For example, 4×3, $a^2 b^3$, and $\frac{4}{3}\pi r^3$ are all monomials.

MONTE CARLO SIMULATION A Monte Carlo simulation uses a random number generator to model a series of events. This method is used when it is uncertain whether or not a particular event will occur, but the probability of occurrence can be estimated. For example, the Monte Carlo method can simulate a baseball game if you know the probability that each player will get a hit. A computer can be programmed to generate a random number for each at bat, and then determine whether or not a hit occurred.

MULTINOMIAL A multinomial is the sum of two or more monomials. Each monomial is called a term. For example, $a^2b^3 + 6 + 4b^5$ is a multinomial with three terms.

MULTIPLE REGRESSION Suppose that a dependent variable Y depends on some independent variables X_1, X_2, and X_3 according to the equation:

$$Y = \beta_1 X_1 + \beta_2 X_2 + \beta_3 X_3 + \beta_4 + \epsilon$$

where $\beta_1, \beta_2, \beta_3$, and β_4 are unknown coefficients, and ϵ is a random variable called the error term. See **regression** for a discussion of the case where there is only one independent variable. The problem in multiple regression is to use observed values of the X's and Y to estimate the values of the β's. For example, Y could be the amount of money spent on food, X_1 could be income, X_2 could be the price of food, and X_3 could be the average price of other goods. The random variable ϵ is included to account for all other factors that could affect demand for food that are not explicitly listed in the equation. If our equation is going to be of much help in predicting the demand for food, then the factors we have included must be more important than the ones left out.

If we have t observations each for Y, X_1, X_2, and X_3 then we can arrange the observations of the X's into a matrix **X** of t rows and four columns (with the last column consisting only of ones). **Y** can be arranged into a matrix of t rows and one column. If the coefficients are arranged in a matrix of four rows and one column β, the estimate for the coefficients is:

$$\beta = (\mathbf{X^{tr}X})^{-1}\mathbf{X^{tr}Y}$$

where $(\mathbf{X}^{tr}\mathbf{X})^{-1}$ is the inverse of the matrix formed by multiplying $\mathbf{X}$ transpose by $\mathbf{X}$. (See **matrix; matrix multiplication.**) This is called the ordinary least squares estimate because it minimizes the squares of the deviations between the actual values of Y and the values of Y predicted by the regression equation. The actual calculations of the regression coefficients are best left to a computer.

The R^2 statistic provides a way of determining how much of the variance in Y this equation is able to explain. The t-statistic for each coefficient provides an estimate of whether that coefficient really should be included in the regression (i.e., is it really different from zero?). Regression methods are used often in statistics and in the branch of economics known as econometrics.

MULTIPLICAND In the equation $ab = c$, a and b are the multiplicands.

MULTIPLICATION Multiplication is the operation of repeated addition. For example, $3 \times 5 = 5+5+5 = 15$. Multiplication is symbolized by a multiplication sign ("$\times$") or by a dot ("$\cdot$"). In algebra much writing can be saved by leaving out the multiplication sign when two letters are being multiplied, or when a number multiplies a letter. For example, the expressions ab, πr^2, and $\frac{1}{2}at^2$ mean $a \times b$, $\pi \times r^2$, and $\frac{1}{2} \times a \times t^2$, respectively.

Multiplication obeys the commutative property:

$$(a \times b) = (b \times a)$$

and the associative property:

$$(a \times b) \times c = a \times (b \times c)$$

Whenever an expression contains both additions and multiplications, the multiplications are done first (unless a set of parentheses indicates otherwise). For example:

$$3 \times 5 + 4 \times 5 \;=\; 15 + 20 = 35$$
$$3 \times (5 + 6) \times 4 \;=\; 3 \times 11 \times 4 = 132$$

The relation between addition and multiplication is given by the distributive property:

$$a(b + c) = ab + ac$$

MULTIPLICATIVE IDENTITY The number 1 is the multiplicative identity, because $1 \times a = a$, for all a.

MULTIPLICATIVE INVERSE The multiplicative inverse of a number a (written as $1/a$ or a^{-1}) is the number that, when multiplied by a, gives a result of 1:

$$a \times \frac{1}{a} = 1$$

The multiplicative inverse is also called the reciprocal. For example, $\frac{1}{2}$ is the reciprocal of 2. There exists a multiplicative inverse for every real number except zero.

N

NAPIER John Napier (1550 to 1617) was a Scottish mathematician who developed the concept of logarithms.

NATURAL LOGARITHM The natural logarithm of a positive number x (written as $\ln x$) is the logarithm of x to the base e, where $e = 2.71828\ldots$. The natural logarithm function can also be defined by the definite integral

$$\ln x = \int_1^x t^{-1} dt$$

(See figure 70.)

Here is a table of some natural logarithms:

x	$\ln x$	x	$\ln x$
0.2	−1.6094	5	1.6094
0.5	−0.6931	6	1.7918
0.8	−0.2231	7	1.9459
1	0	8	2.0794
2	0.6931	9	2.1972
3	1.0986	10	2.3026
4	1.3863	100	4.6052

NATURAL NUMBERS The natural numbers are the set of numbers $\{1, 2, 3, 4, 5, 6, 7, 8, \ldots\}$. This set of numbers is also called the counting numbers, since they're the numbers used to count something. They can also be called the positive integers.

NECESSARY In the statement $p \rightarrow q$, q is a necessary condition for p to be true. For example, having

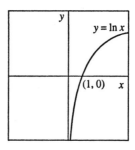

Figure 70

four 90° angles is a necessary condition for a quadrilateral to be a square (but it is not a sufficient condition).

NEGATION The negation of a statement p is the
statement NOT p. (See **logic; Boolean algebra**.)

NEGATIVE A negative number is any real number
less than zero. The negative of any number a (written as $-a$) is defined by this equation: $a + (-a) = 0$.
  These are the rules for operations with negative
numbers:
  1. To add two negative numbers: Add the two
absolute values, and give the result a negative sign.
Example: $(-5) + (-3) = -8$.
  2. To add one positive and one negative number:
Subtract the two absolute values, giving the result
a positive sign if the positive number had greater
absolute value, and giving the result a negative sign
if the negative number had greater absolute value.
Examples: $5 + (-3) = 2$; $(-5) + 3 = -2$.
  3. To multiply two negative numbers: multiply

the two absolute values and give the result a positive
sign. Example: $(-5) \times (-3) = 15$.

4.   To multiply one positive and one negative
number: multiply the two absolute values and give
the result a negative sign. Example: $(-5) \times (3) =$
$(5) \times (-3) = -15$.

5. To take the square root of a negative number,
see **imaginary number.**

NEWTON Sir Isaac Newton (1643 to 1727) was an
English mathematician and scientist who developed
the theory of gravitation and the laws of motion, designed a reflecting telescope using a paraboloid mirror, used a prism to split white light into component
colors, and was one of the inventors of calculus (independently of his rival Leibniz). (See **Newton's
method.**)

NEWTON'S METHOD Newton's method (see figure 71) provides a way to estimate the places where
complicated functions cross the x-axis. First, make
a guess, x_1, that seems reasonably close to the true
value. Then approximate the curve by its tangent
line to estimate a new value, x_2, from the equation

$$x_2 = x_1 - \frac{f(x_1)}{f'(x_1)}$$

where $f'(x_1)$ is the derivative of the function f at
the point x_1. (See **calculus; derivative.**)

The process is iterative; that is, it can be repeated
as often as you like. This means that you can get as
close to the true value as you wish.

For example, Newton's method can be used to
find the x-intercept of the function $f(x) = x^3 - 2x^2 -$

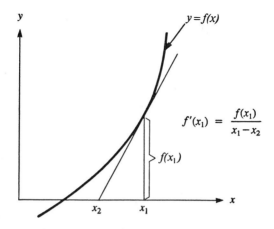

Figure 71   Newton's method

$6x - 8$, whose derivative is $f'(x) = 3x^2 - 4x - 6$. Start with a guess, $x_1 = 10$:

x_i	$f(x_i)$	$f'(x_i)$	$-f(x)/f'(x)$
10	732	254	-2.88
7.118	209	117	-1.77
5.343	55.4	58.3	-0.95
4.393	11.8	34.3	-0.34
4.048	1.28	26.98	-0.047
4.00088	0.023		

The true value of the intercept is $x = 4$.

A brief word of warning: the method doesn't always work. The tangent line approximation will not always converge to the true value. The method will not work for the function shown in figure 72.

NOETHER   Emmy Noether (1882 to 1935) was a German mathematician who contributed to abstract algebra.

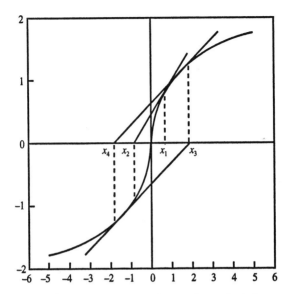

Figure 72 Function where Newton's method does not converge

NON-EUCLIDIAN GEOMETRY Euclidian geometry describes the geometry of our everyday world. One postulate of Euclidian geometry describes the behavior of parallel lines. This postulate says that, if a straight line crosses two coplanar straight lines, and if the sum of the two interior angles formed on one side of the crossing line is less than 180°, then the two other lines will intersect at some point. In other words, they will not be parallel. If, on the other hand, the sum of the two interior angles is 180°, then the two lines will be parallel, meaning that they could be extended forever and never intersect. This postulate seems intuitively clear, but nobody has been able to prove it after several centuries of trying. Since we cannot travel to infinity to verify that two seemingly parallel lines never intersect, we cannot tell whether this postulate really is satisfied in our universe.

Some mathematicians decided to investigate what would happen to geometry if they changed the parallel postulate. They found that they were able to prove theorems in their new type of geometry. These theorems were consistent because no two theorems contradicted each other, but the geometry that resulted was different from the geometry developed by Euclid. In one type of non-Euclidean geometry, called hyperbolic geometry, there is more than one line parallel to a given line through a given point. Janos Bolyai wrote one of the earliest descriptions of hyperbolic geometry in 1823; Nicolai Lobachevsky independently developed the same ideas at the same time. In another type of non-Euclidean geometry, called elliptic geometry, there are no parallel lines. Elliptic geometry generalizes the situation in which you would find yourself if you were a two-dimensional being confined to the surface of a sphere. In that case any two "lines" would always intersect on the other side of the sphere. Ludwig Schlafli and Bernhard Riemann described elliptic geometry in the late 1800s.

Non-Euclidean geometries play an important role in the development of relativity theory. They also are important because they shed light on the nature of logical systems.

NORMAL In mathematics the word "normal" means
"perpendicular." A line is normal to a curve if it is
perpendicular to a tangent line to that curve at the
point where it intersects the curve. Two vectors are
normal to each other if their dot product is zero.

NORMAL DISTRIBUTION A continuous random
variable X has a normal distribution if its density
function is

$$f(x) = \frac{1}{\sigma\sqrt{2\pi}}e^{-(x-\mu)^2/2\sigma^2}$$

The mean (or expectation) of X is μ, and its
variance is σ^2. If $\mu = 0$ and $\sigma = 1$, then X is said
to have the standard normal distribution, which has
the density function

$$f(x) = \frac{1}{\sqrt{2\pi}}e^{-x^2/2}$$

Figure 73 shows a graph of the standard normal
density function. There is no formula for this integral, but Tables 3 and 4 list some values.

Also, the value of the integral can be found from
this Taylor series:

$$\frac{1}{\sqrt{2\pi}}\int_0^x e^{-.5t^2}\,dt$$

$$= \frac{1}{\sqrt{2\pi}}\left[x - \frac{x^3}{6} + \frac{x^5}{40} - \frac{x^7}{336} + \frac{x^9}{3456} - \frac{x^{11}}{42240}\cdots\right]$$

If the first x in the series is called term 0, then
the denominator in term i is found from the formula
$2^i i!(2i+1)$.

The central limit theorem is one important application of the normal distribution. The central limit

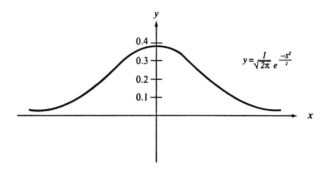

Figure 73 Density function—Standard Normal Random Variable

theorem states that, if $X_1, X_2, \ldots, X_n$ are independent, identically distributed random variables, each with mean μ and variance σ^2, then, in the limit that n goes to infinity,

$$S_n = X_1 + X_2 + X_3 + \cdots + X_n$$

will have a normal distribution with mean $n\mu$ and variance $n\sigma^2$. The reason that this theorem is so remarkable is that it is completely general. It says that, no matter how X is distributed, if you add up enough measurements, the sum of the X's will have a normal distribution.

NOT The word "NOT" is used in logic to indicate the negation of a statement. The statement "NOT p" is false if p is true, and it is true if p is false. The operation of NOT can be described by this truth table:

p	NOT p
T	F
F	T

The symbols $\neg p$ or $\sim p$ or $\bar{p}$ are used to represent NOT. (See **logic; Boolean algebra.**)

NULL HYPOTHESIS The null hypothesis is the hypothesis that is being tested in a hypothesis-testing situation. (See **hypothesis testing.**) Often the null hypothesis is of the form "There is no relation between two quantities." For example, if you were testing the effect of a new medicine, you would want to test the null hypothesis "This medicine has no effect on the patients who take it." If the medicine did work, then you would obtain statistical evidence that would cause you to reject the null hypothesis.

NULL SET The null set is the set that contains no elements. The term "null set" means the same as the term "empty set."

NUMBER Everyone first learns the basic set of numbers: 1, 2, 3, 4, 5, 6, ... These are known as the natural numbers, or counting numbers. The natural numbers are used to count discrete objects, such as two books, five trees, or five thousand people. There is an infinite number of natural numbers. Natural numbers obey an important property known as closure under addition. This means that, whenever you add two natural numbers together, the result will still be a natural number. The natural numbers also obey closure under multiplication.

One important number not included in the set of natural numbers is zero. It would be very difficult to measure the snowfall in the Sahara Desert without knowing the number zero. The union of the set of natural numbers and the set containing zero is the set of whole numbers.

The set of whole numbers does not obey closure under subtraction. If you subtract one whole number from another, there is no guarantee that you will get another whole number. This suggests the need for another kind of number: negative numbers. Also, there are times when the natural numbers do not do an adequate job of measuring certain quantities. If you are measuring the government surplus (equal to tax revenue minus government expenditures), you need negative numbers to represent the years when the government runs a deficit. If you are measuring the yardage gained by a football team, you need to use negative numbers to represent the yardage on the plays when the team loses yardage. Every natural number has its own negative, or additive inverse. If a represents a natural number and $-a$ is its negative, then $a + (-a) = 0$. The union of the set of natural numbers and the set of the negatives of all the natural numbers and zero is the set of integers. The set of integers looks like this:

$$\ldots, -5, -4, -3, -2, -1, 0, 1, 2, 3, 4, 5, 6, 7, 8, \ldots$$

Integers do not obey closure under division. A rational number is any number that can be obtained as the result of a division problem containing two integers. All fractions, such as $\frac{1}{2}$, 0.6, 3.4, and $5\frac{2}{3}$, are rational numbers. Also, all the integers are rational numbers, since any integer a can be written as $a/1$. The set of rational numbers is infinitely dense because there is always an infinite number of other rational numbers between any two rational numbers.

Nevertheless, there are many numbers that aren't rational. The square roots of most integers are not rational. For example, $\sqrt{4} = 2$, but $\sqrt{5}$ is approximately equal to 2.236067977..., which cannot be expressed as the ratio of two integers. There are important geometric reasons for needing these irrational numbers. (See **Pythagorean theorem**.) Irrational numbers are also needed to express most of the values for trigonometric functions, and two special numbers, pi = π = 3.14159... and e = 2.71828... are both irrational. For practical purposes you can always find a rational number that is a close approximation to any irrational number.

The set of all rational numbers and all irrational numbers is known as the set of real numbers. Each real number can be represented by a unique point on a straight line that extends off to infinity in both directions. Real numbers have a definite order, that is, for any two distinct real numbers you can always tell which one is bigger. The result of a measurement of a physical quantity, such as energy, distance, or momentum, will be a real number.

However, there are some numbers that are not real. There is no real number x that satisfies the equation $x^2 + 1 = 0$. Imaginary numbers are needed to describe the square roots of negative numbers. The basis of the imaginary numbers is the imaginary unit, i, which is defined so that $i^2 = -1$. Pure imaginary numbers are formed by multiplying a real number by i. For example: $\sqrt{-64} = \sqrt{64}\sqrt{-1} = 8i$

If a pure imaginary number is added to a real number, the result is known as a complex number. The real numbers and the imaginary numbers are both subsets of the set of complex numbers. The general form of a complex number is $a + bi$, where a and b are both real numbers. Complex numbers are important in some areas of physics.

NUMBER LINE A number line is a line on which each point represents a real number. (See **real number**.)

NUMBER THEORY Number theory is the study of properties of the natural numbers. One aspect of number theory focuses on prime numbers. For example, it can be easily proved that there are an infinite number of prime numbers. Suppose, for example, that p was the largest prime number. Then, form a new number equal to one plus the product of all the prime numbers from 2 up to p. This number will not be divisible by any of these prime numbers (and, therefore, not by any composite number formed by multiplying these primes together) and will therefore be prime. This contradicts the assumption that p is the largest prime number. There are still unsolved problems involving the frequency of occurrence of prime numbers.

The introduction of computers has made it possible to verify that a proposition works for very large numbers, but no computer can count all the way to inifinity so the computer is no subsitute for a formal proof if you need to know that a theorem is always true.

For another example of a problem in number theory, see **Fermat's last theorem**.

NUMERAL A numeral is a symbol that stands for a number. For example, "4" is the Arabic numeral for the number four. "IV" is the Roman numeral for the same number.

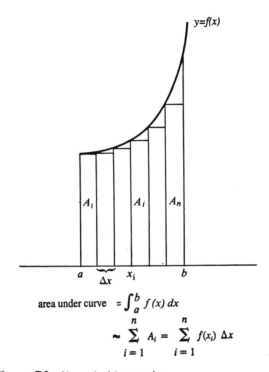

area under curve $= \int_a^b f(x)\,dx$

$$\sim \sum_{i=1}^{n} A_i = \sum_{i=1}^{n} f(x_i)\,\Delta x$$

Figure 74 Numerical integration

NUMERATOR The numerator is the number above the bar in a fraction. In the fraction $\frac{8}{9}$, 8 is the numerator. (See **denominator; fraction**.)

NUMERICAL INTEGRATION The numerical integration method is used when it is not possible to find a formula that can be evaluated to give the value of a definite integral. For example, there is no formula that gives the value of the definite integral

$$\int_0^a e^{-x^2} dx$$

The procedure in numerical integration is to divide the area under the curve into a series of tiny rectangles and then add up the areas of the rectangles. (See figure 74.) The height of each rectangle is equal to the value of the function at that point. As the number of rectangles increases (and the width of each rectangle becomes smaller), the accuracy of the method improves. In practice, the calculations for a numerical integration are carried out by a computer. There are also alternative methods that use trapezoids or strips bounded by parabolas.

O

OBJECTIVE FUNCTION An objective function is a function whose value you are trying to maximize or minimize. The value of the objective function depends on the values of a set of choice variables, and the problem is to find the optimal values for those choice variables. For an example, see **linear programming**.

OBLATE SPHEROID An oblate spheroid is elongated horizontally. For contrast, see **prolate spheroid**.

OBLIQUE ANGLE An oblique angle is an angle that is not a right angle.

OBLIQUE TRIANGLE An oblique triangle is a triangle that is not a right triangle.

OBTUSE ANGLE An obtuse angle is an angle larger than a 90° angle and smaller than a 180° angle.

OBTUSE TRIANGLE An obtuse triangle (see figure 75) is a triangle containing one obtuse angle. (Note that a triangle can never contain more than one obtuse angle.)

Figure 75 Obtuse triangles

Figure 76   Octahedron

OCTAGON An octagon is an eight-sided polygon. The best-known example of an octagon is a stop sign. (See **polygon**.)

OCTAHEDRON An octahedron is a polyhedron with eight faces. (See **polyhedron**.) (See figure 76.)

OCTAL An octal number system is a base-eight number system.

ODD FUNCTION The function $f(x)$ is an odd function if it satisfies the property that $f(-x) = -f(x)$. For example, $f(x) = \sin x$ and $f(x) = x^3$ are both odd functions. For contrast, see **even function**.

ODD NUMBER An odd number is a whole number that is not divisible by 2, such as 1, 3, 5, 7, 9, 11, 13, 15, ... For contrast, see **even number**.

ONE-TAILED TEST In a one-tailed test the critical region consists of only one tail of a distribution. The null hypothesis is rejected only if the test statistic has an extreme value in one direction. (See **hypothesis testing**.)

OPEN INTERVAL An open interval is an interval that does not contain both its endpoints. For example, the interval $0 < x < 1$ is an open interval because the endpoints 0 and 1 are not included. For contrast, see **closed interval**.

OPEN SENTENCE An open sentence is a sentence containing one or more variables that can be either true or false, depending on the value of the variable(s). For example, $x = 7$ is an open sentence.

OPERAND An operand is a number that is the subject of an operation. In the equation $5 + 3 = 8$, 5 and 3 are the operands.

OPERATION An operation, such as addition or multiplication, is the process of carrying out a particular rule on a set of numbers. The four fundamental arithmetic operations are addition, multiplication, division, and subtraction.

OPTICS Optics (also known as geometric optics) is the study of how light rays behave when they are reflected or bent by various media. In particular, optics focuses on light rays that are reflected off mirrors, or are refracted (bent) by lenses.

A reflecting telescope is built by taking advantage of the fact that parallel light rays striking a parabolic mirror will all be reflected back to the focal point. (See figure 77.) (See **parabola; angle of incidence**.)

A refracting telescope is built by designing a lens that will refract parallel light rays to a single point. (See figure 78.) (See **Snell's law**.)

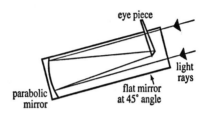

Figure 77   Reflecting Telescope

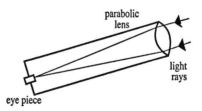

Figure 78   Refracting Telescope

OR The word "OR" is a connective word used in logic. The sentence "*p* OR *q*" is false only if both *p* and *q* are false; it is true if either *p* or *q* or both are true. The operation of OR is illustrated by the truth table:

p	*q*	*p* OR *q*
T	T	T
T	F	T
F	T	T
F	F	F

The symbol $\vee$ is often used to represent OR. An OR sentence is also called a *disjunction*. (See **logic; Boolean algebra**.)

ORDERED PAIR An ordered pair is a set of two numbers where the order in which the numbers are written has an agreed-upon meaning. One common example of an ordered pair is the Cartesian coordinates (x, y), where it is agreed that the horizontal coordinate is always listed first and the vertical coordinate last.

ORDINATE The ordinate of a point is another name for the y coordinate. (See **Cartesian coordinates; abscissa**.)

ORIGIN The origin is the point $(0, 0)$ in Cartesian coordinates. It is the point where the x- and the y-axes intersect.

ORTHOCENTER The orthocenter of a triangle is the point where the three altitudes of the triangle meet. (See **triangle**.)

P

PARABOLA A parabola (see figure 79) is the set of
all points in a plane that are equally distant from a
fixed point (called the *focus*) and a fixed line (called
the *directrix*). If the focus is at $(0, a)$ and the directrix is the line $y = -a$, then the equation can be
found from the definition of the parabola:

$$
\begin{aligned}
y + a &= \sqrt{x^2 + (y - a)^2} \\
y^2 + 2ay + a^2 &= x^2 + y^2 - 2ay + a^2 \\
4ay &= x^2 \\
y &= \frac{1}{4a}x^2
\end{aligned}
$$

The final equation for a parabola is very simple.
One example of a parabola is the graph of the equation $y = x^2$.

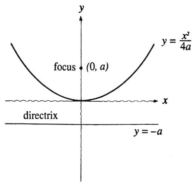

Figure 79   Parabola

Parabolas have many practical uses. The course of a thrown object, such as a baseball, is a parabola (although it will be modified a bit by air resistance). The cross section of a telescopic mirror is a parabola. The telescopic mirror constitutes a surface known as a paraboloid, which is formed by rotating a parabola about its axis. When parallel light rays from a distant star strike the paraboloid, they are all reflected back to the focal point. (See **optics**.) For the same reason, the network microphones that pick up field noises at televised football games are shaped like paraboloids. Probably the largest parabola in practical use is the cross section of the 1000-foot-wide radio telescope carved out of the ground at Arecibo, Puerto Rico. The parabola is an example of a more general class of curves known as **conic sections**.

PARABOLOID A paraboloid is a surface that is formed by rotating a parabola about its axis. (See **parabola**.)

PARALLEL Two lines are parallel if they are in the same plane but never intersect. In figure 80 lines *AB* and *CD* are parallel. A postulate of Euclidian geometry states that "Through any point not on a line there is one and only one line that is parallel to the first line."

Two planes are parallel if they never intersect.

PARALLELEPIPED A parallelepiped is a solid figure with six faces such that the planes containing two opposite faces are parallel. (See figure 81.) Each face is a parallelogram.

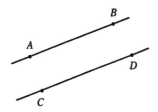

Figure 80   Parallel lines

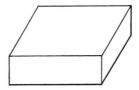

Figure 81   Parallelepiped

PARALLELOGRAM A parallelogram is a quadrilateral with opposite sides parallel. (See **quadrilateral**.)

PARAMETER (1) In statistics a parameter is a quantity (often unknown) that characterizes a population. For example, the mean height of all 6-year-olds in the United States is an unknown parameter. One of the goals of statistical inference is to estimate the values of parameters.

(2) See **parametric equation.**

PARAMETRIC EQUATION A parametric equation in x and y is an equation of the form $x = f(t), y = g(t)$, where t is the parameter, and f and g are two functions. For example, the parametric equation $x = r \cos t$, $y = r \sin t$ defines the circle centered at the origin with radius r. For another example of a parametric equation, see **cycloid**.

PARENTHESIS A set of parentheses ( ) indicates that the operation in the parentheses is to be done first. For example, in the expression

$$y = 5 \times (2 + 10 + 30) = 5 \times 42 = 210$$

the parentheses tell you to do the addition first.

PARTIAL DERIVATIVE The partial derivative of $y = f(x_1, x_2, \ldots, x_n)$ with respect to x_i is found by taking the derivative of y with respect to x_i, while all the other independent variables are held constant. (See **derivative**.) For example, suppose that y is this function of two variables: $y = x_1^a x_2^b$. Then the partial derivative of y with respect to x_1 (written as $\partial y/\partial x_1$) is $ax_1^{a-1}x_2^b$. Likewise, the partial derivative with respect to x_2 is found by taking the derivative with x_1 treated as a constant:

$$\frac{\partial y}{\partial x_2} = bx_1^a x_2^{b-1}$$

PARTIAL FRACTIONS An algebraic expression of the form

$$\frac{b_m x^m + b_{m-1} x^{m-1} + \cdots + b_2 x^2 + b_1 x + b_0}{(x - a_1)(x - a_2)(x - a_3) \times \cdots \times (x - a_{n-1})(x - a_n)}$$

(where $m < n$) can be written as the sum of n partial fractions, like this:

$$\frac{C_1}{x - a_1} + \frac{C_2}{x - a_2} + \cdots + \frac{C_n}{x - a_n}$$

where $C_1, \ldots C_n$, are constants for which we can solve.

For example, the expression

$$\frac{5x - 7}{(x - 1)(x - 2)}$$

can be split up into partial fractions as follows:

$$\frac{5x - 7}{(x - 1)(x - 2)} = \frac{C_1}{x - 1} + \frac{C_2}{x - 2}$$

Now we need to solve for C_1 and C_2, which we can do this way:

$$\frac{5x - 7}{(x - 1)(x - 2)} = \frac{C_1(x - 2) + C_2(x - 1)}{(x - 1)(x - 2)}$$

For this equation to be true for all values of x, we must have C_1 and C_2 satisfy these two equations:

coefficients of x:

$$5 = C_1 + C_2$$

constant terms:

$$-7 = -2C_1 - C_2$$

This is a two-equation, two-unknown system, which has the solution $C_1 = 2, C_2 = 3$. Therefore:

$$\frac{5x - 7}{(x - 1)(x - 2)} = \frac{2}{x - 1} + \frac{3}{x - 2}$$

PASCAL Blaise Pascal (1623 to 1662) was a French mathematician who developed the modern theory of probability, invented a calculating machine using wheels to represent numbers, studied fluid pressure, and wrote about religion. (See **Pascal's triangle**.)

PASCAL'S TRIANGLE Pascal's triangle is a triangular array of numbers in which each number is equal to the sum of the two numbers above it (one is above and left, the other is above and right). Diagonal lines of 1's make up the top two sides of the triangle, which looks like this:

```
                    1
                 1     1
              1     2     1
           1     3     3     1
        1     4     6     4     1
     1     5    10    10     5     1
   1    6    15    20    15     6     1
 1    7    21    35    35    21     7    1
1   8   28    56    70    56    28     8  1
1  9   36   84   126   126   84    36    9  1
```

If the "1" at the top is called row zero, and the first item in each row is called item 0, then item j in row n can be found from the formula:

$$\binom{n}{j} = \frac{n!}{(n-j)!j!}$$

(See **factorial; combinations.**)

Also, row n of the triangle gives the coefficients of the expansion of $(a+b)^n$. (See **binomial theorem.**)

PENTAGON A pentagon is a five-sided polygon. For picture, see **polygon.** The sum of the angles in a pentagon is 540°. A regular pentagon has all five sides equal, and each of the five angles equal to 108°. The most famous pentagon is the Pentagon building, near Washington, D.C., which has sides 921 feet long.

PERCENT A percent is a fraction in which the denominator is assumed to be 100. The symbol % means "percent." For example, 50% means 50/100

= 0.50, 2% means 2/100 = 0.02 and 150% means 150/100 = 1.5.

PERCENTILE The pth percentile of a list is the number such that p percent of the elements in the list are less than that number. For example, if the height of a particular child is at the 55th percentile, then 55 percent of the children of the same age have heights less than this child.

PERFECT NUMBER A perfect number equals the sum of all its factors except itself. For example, the factors of 6 are 1, 2, 3, and 6; since $1 + 2 + 3 = 6$, 6 is a perfect number.

PERIMETER The perimeter of a polygon is the sum of the lengths of all the sides. If you had to walk all the way around the outer edge of a polygon, the total distance you would walk would be the perimeter.

PERIOD The period of a periodic function is a measure of how often the function repeats the same values. For example, the function $f(x) = \cos x$ repeats its values every 2π units, so its period is 2π.

PERIODIC A periodic function is a function that keeps repeating the same values. Formally, a function $f(x)$ is periodic if there exists a number p such that $f(x + p) = f(x)$, for all x. If p is the smallest number with this property, then p is called the period. For example, the function $y = \sin x$ is a periodic function with a period of 2π, because $\sin(x + 2\pi) = \sin x$, for all x. (See **Fourier series**.)

PERMUTATIONS The term "permutations" refers to the number of different ways of choosing things from a group of n objects, when you care about the order in which they are chosen, and the selection is made without replacement. The number of permutations of n objects, taken j at a time, is $n!/(n-j)!$. (See **factorial**.) For example, if there are 25 players on a baseball team, then the total number of possible batting orders is

$$25 \cdot 24 \cdot 23 \cdot 22 \cdot 21 \cdot 20 \cdot 19 \cdot 18 \cdot 17 = \frac{25!}{(25-9)!} = 7.41 \times 10^{11}$$

There are 25 choices for the first batter. Once the first batter is chosen, then there are 24 choices left for the second batter. Once these choices have been made, there are 23 choices left for the second batter, and so on.

For situations where you do not care about the order in which the objects are selected, see **combinations.**

PERPENDICULAR Two lines are perpendicular if the angle between them is a 90° angle. By definition, the two legs of a right triangle are perpendicular to each other. (See figure 82.)

Two vectors are perpendicular if their dot product is zero. (See **dot product.**)

Two planes are perpendicular if the dihedral angle they form is a right angle. (See **dihedral angle.**) In a well-designed house the walls are perpendicular to the floor.

PI The Greek letter π (pi) is used to represent the ratio between the circumference of a circle and its diameter:

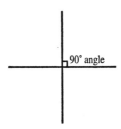

Figure 82   Perpendicular lines

$$\pi = \frac{circumference}{diameter}$$

This ratio is the same for any circle. π is an irrational number with the decimal approximation 3.1415926536 ... π can also be approximated by the fraction 22/7, or 377/120. For example, if a circle has a radius of 8 units, then it has a diameter of 16, a circumference of $16\pi \approx 16 \times 22/7 = 50.3$, and an area of $\pi r^2 = \pi \times 8^2 \approx 201.1$.

There are several ways to find numerical approximations for pi. If we inscribe a regular polygon inside a circle (see figure 83), then the perimeter of the polygon is less than the circumference of the circle. However, if we double the number of sides in the polygon, keeping it inscribed in the same circle, then the perimeter of the polygon will be a closer approximation to the circumference of the circle. If we keep doubling the number of sides, we can come as close as we want to the true circumference. Let s_n be the length of a side of a regular n-sided polygon inscribed in a circle of radius r. Then:

$$s_{2n}^2 = 2r^2 - r\sqrt{4r^2 - s_n^2}$$

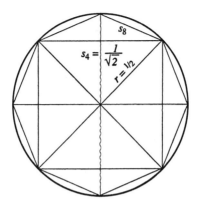

Figure 83 Approximating pi by the perimeter of a polygon

For $r = \frac{1}{2}$, the perimeter of the polygon will approach π as the number of sides is increased:

n	nS_n (approximation for π)
4	2.8284
8	3.0615
16	3.1214
32	3.1365
64	3.1403
128	3.1413
256	3.1415
512	3.14157
1024	3.14159

The arctangent function can be used to find a series approximation for π. We know that

$$\frac{1}{1-z} = 1 + z + z^2 + z^3 + z^4 + \ldots$$

(for $|z| < 1$). (See **geometric series.**) If $z = -x^2$, then

$$\frac{1}{1+x^2} = 1 - x^2 + x^4 - x^6 + x^8 - x^{10} + \cdots$$

If $y = \arctan x$, then $dy/dx = 1/(1+x^2)$. (See **integral.**) Then:

$$\frac{dy}{dx} = 1 - x^2 + x^4 - x^6 + x^8 - x^{10} + \cdots$$

If we integrate this series term by term, then

$$y = \arctan x = x - \frac{x^3}{3} + \frac{x^5}{5} - \frac{x^7}{7} + \frac{x^9}{9} - \cdots$$

Since $\tan(\pi/4) = 1$, then $\arctan 1 = \pi/4$. Therefore:

$$\frac{\pi}{4} = 1 - \frac{1}{3} + \frac{1}{5} - \frac{1}{7} + \frac{1}{9} - \frac{1}{11} + \cdots$$

After 1,000 terms, this series gives the value 3.1406; after 1,001 terms, the result is 3.1426.

Another way to find π is the infinite product:

$$\frac{\pi}{2} = \frac{2}{1} \times \frac{2}{3} \times \frac{4}{3} \times \frac{4}{5} \times \frac{6}{5} \times \frac{6}{7} \times \frac{8}{7} \times \frac{8}{9} \times \cdots$$

PIECEWISE A function is piecewise continuous if it can be broken into different segments such that it is continuous in each segment.

PLACEHOLDER Zero acts as a placeholder to indicate which power of 10 a digit is to be multiplied by. The importance of this role is indicated by considering the difference between the two numbers $300 = 3 \times 10^2$ and $3,000,000 = 3 \times 10^6$.

PLANE A plane is a flat surface (like a tabletop) that stretches off to infinity. A plane has zero thickness, but infinite length and width. "Plane" is one of the key undefined terms in Euclidian geometry Any three noncollinear points will determine one and only one plane.

PLATO Plato (428 BC to 348 BC), one of the greatest of ancient Greek philosophers, established the Academy at Athens with these words over the entrance: "Let no one ignorant of geometry enter here." The five regular polyhedra are sometimes called Platonic solids.

POISSON Simeon-Denis Poisson (1781 to 1840) was a French mathematician who made contributions to celestial mechanics, probability theory, and the theory of electricity and magnetisim. (See **Poisson distribution.**)

POISSON DISTRIBUTION The Poisson distribution is a discrete random variable distribution that often describes the frequency of occurrence of certain random events, such as the number of phone calls that arrive at an office in an hour. The Poisson distribution can also be used as an approximation for the binomial distribution. The Poisson distribution is characterized by a parameter usually written as λ (the Greek letter lambda). If X has a Poisson distribution, then the probability function is given by the formula:

$$\Pr(X = k) = \frac{e^{-\lambda}\lambda^k}{k!}$$

where $e = 2.71828\ldots$, and the exclamation mark indicates factorial. The Poisson distribution has the unusual property that the expectation and the variance are equal (each is equal to λ).

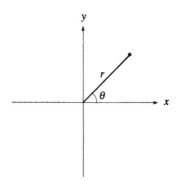

Figure 84 Polar coordinates

POLAR COORDINATES Any point in a plane can
be identified by its distance from the origin (r) and
its angle of inclination (θ). This type of coordinate
system is called a polar coordinate system. (See figure 84.) It is an alternative to rectangular (Cartesian) coordinates. Polar coordinates can be changed
into Cartesian coordinates by the formulas

$$x = r\cos\theta, \; y = r\sin\theta$$

Rectangular coordinates can be changed into polar coordinates by the formulas

$$r = \sqrt{x^2 + y^2}, \; \theta = \arctan\frac{y}{x}$$

For example:

Cartesian Coordinates	Polar Coordinates
$(3,0)$	$(3, 0°)$
$(0,4)$	$(4, 90°)$
$(\sqrt{3}, 1)$	$(2, 30°)$
$(3, 3)$	$(\sqrt{18}, 45°)$

The equation of a circle in polar coordinates is very simple: $r = R$, where R is the radius. The formula for the rotation of axes in polar coordinates is also very simple: $r' = r$, $\theta' = \theta - \phi$, where ϕ is the angle of rotation. (See **rotation**.)

POLISH NOTATION In Polish notation, operators are written before their operands. Thus, $3 + 5$ is written $+\ 5\ 3$. No parentheses are needed when this notation is used. For example, $(2 + 3) \times 4$ would be written $\times + 2\ 3\ 4$.

POLYGON A polygon (see figure 85) is the union of several line segments that are joined end to end so as to completely enclose an area. "Polygon" means "many-sided figure." Most useful polygons are convex polygons; in other words, the line segment connecting any two points inside the polygon will always stay completely inside the polygon. (A polygon that is not convex is concave, that is, it is caved in.)

Polygons are classified by the number of sides they have. The most important ones are triangles (three sides), quadrilaterals (four sides), pentagons (five sides), hexagons (six sides), and octagons (eight sides). A polygon is a *regular polygon* if all its sides and angles are equal.

Two polygons are congruent (figure 86) if they have exactly the same shape and size. Two polygons are similar if they have exactly the same shape but different sizes. Corresponding angles of similar polygons are equal and corresponding sides have the same ratio.

The sum of all the angles in a polygon with n sides is $(n - 2)180°$.

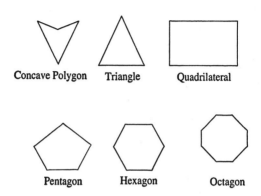

Figure 85 Polygons

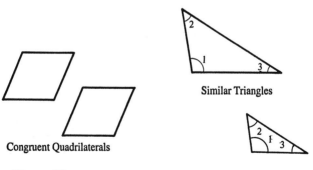

Figure 86

POLYHEDRON A polyhedron is a solid that is bounded by plane polygons. The polygons are called the faces; the lines where the faces intersect are called the edges; and the points where three or more faces intersect are called the vertices. Some examples of polyhedrons are cubes, tetrahedrons, pyramids, and prisms.

There are five regular polyhedra, which means that each face is a congruent regular polygon. For pictures, see the entries for each type.

Type	Face Shape	F	V	E
tetrahedron	triangle	4	4	6
cube	square	6	8	12
octahedron	triangle	8	6	12
dodecahedron	pentagon	12	20	30
icosahedron	triangle	20	12	30

(F stands for the number of faces; V is the number of vertices; and E is the number of edges.) Note that the $F + V = E + 2$.

POLYNOMIAL A polynomial in x is an algebraic expression of the form

$$a_n x^n + a_{n-1} x^{n-1} + \cdots + a_3 x^3 + a_2 x^2 + a_1 x + a_0$$

where $a_0, a_1, \cdots, a_n$ are constants that are the coefficients of the polynomial, and n is a positive integer. In this article it will be assumed that all of the coefficients $a_0 \ldots a_n$ are real numbers.

The degree of the polynomial is the highest power of the variable that appears. The polynomial listed above has degree n, the polynomial $x^2 + 2x + 4$ has degree 2, and the polynomial $3y^3 + 2y$ has degree 3.

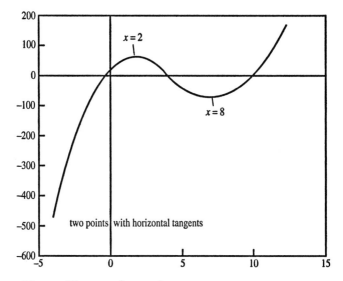

Figure 87 $y = x^3 - 15x^2 + 48x + 12$

Graphs of polynomial functions are interesting because the curve can change directions. The number of turning points is odd if the degree of the polynomial is even, and vice versa, and the maximum number of turning points is one less than the degree of the polynomial. The table shows the number of turning points a polynomial curve might have. Figure 87 shows a third-degree polynomial curve with two turning points.

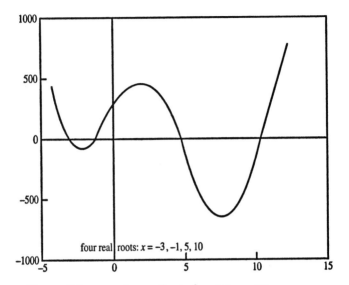

four real roots: $x = -3, -1, 5, 10$

Figure 88 $y = x^4 - 11x^3 - 7x^2 + 155x + 150$

	Number of	
Degree	**turning points**	**Term**
1	0	straight line
2	1	quadratic
3	0 or 2	cubic
4	1 or 3	quartic
5	0 or 2 or 4	quintic

At each turning point the curve has a horizontal tangent line. The value of x at each of these points can be found by setting the derivative of the curve equal to zero. (See **derivative**.)

A polynomial equation is an equation with a poly-

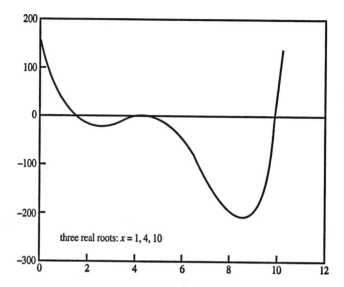

three real roots: $x = 1, 4, 10$

Figure 89 $y = x^4 - 19x^3 + 114x^2 - 256x + 160$

nomial on one side and zero on the other side:

$$a_n x^n + a_{n-1} x^{n-1} + \cdots + a_3 x^3 + a_2 x^2 + a_1 x + a_0 = 0$$

A polynomial of degree n can be written as the product of n first-degree (or linear) factors, so the polynomial equation can be rewritten:

$$(x - r_1)(x - r_2) \times \cdots \times (x - r_n) = 0$$

The equation will be true if either $x = r_1$, or $x = r_2$, and so on, so the equation will have n solutions. In general, a polynomial equation of degree n will have n solutions. However, there are two complications. First, not all of the solutions may be distinct. For example, the equation

$$x^2 - 4x + 4 = (x - 2)(x - 2) = 0$$

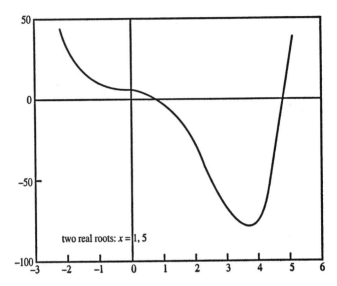

Figure 90 $y = x^4 - 4x^3 - 5x^2 - 2x + 10$

has two solutions, but they are both equal to 2. An extreme example is the equation $(x - a)^n = 0$, which has n solutions, but they are all equal to a.

Second, not all of the solutions will be real numbers. For example, the equation

$$x^2 + 2x + 2 = 0$$

has two solutions: $x = -1 + i$, and $x = -1 - i$. The letter i satisfies $i^2 = -1$. (See **imaginary number**.) These solutions are both said to be **complex numbers**. The complex solutions to a polynomial equation come in pairs: if $(u + vi)$ is a solution to a polynomial equation, then $(u - vi)$ will also be a solution (remember the assumption that all of the

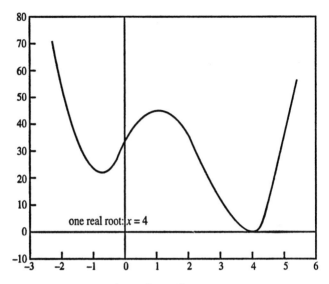

one real root: $x = 4$

Figure 91 $y = x^4 - 6x^3 + 2x^2 + 16x + 32$

coefficients are real numbers).

If the degree of the polynomial equation is two, then the equation is called a **quadratic equation**. These equations can be solved fairly easily. It can be difficult to solve polynomial equations if the degree is higher than two. The **rational root theorem** can sometimes help to identify rational roots. **Newton's method** is a way to find numerical approximations to the roots. If you are able to factor the polynomial, then the solutions will be obvious, but factoring can be very difficult. If you have found one solution of the equation, you can make the equation simpler. If you know that $(x = r)$ is a solution of the polynomial equation $f(x) = 0$, then use **synthetic division** to

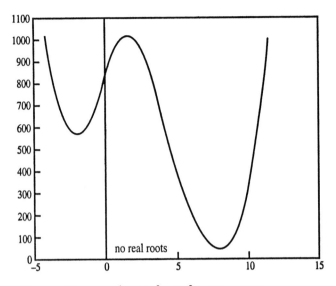

Figure 92 $y = x^4 - 11x^3 - 7x^2 + 155x + 800$

divide $f(x)/(x - r)$. The result will be a polynomial whose degree is one less than $f(x)$, so it will be easier to find more solutions.

Figures 88 to 92 illustrate the different possibilities for a fourth-degree polynomial curve.

POPULATION A population consists of the set of all items of interest. The population may consist of a group of people or some other kind of object. In many practical situations the parameters that characterize the population are unknown. A sample of items is selected from the population, and the characteristics of that sample are used to estimate the characteristics of the population. (See **statistical inference.**)

POSITIVE NUMBER A positive number is any real number greater than zero.

POSTULATE A postulate is a fundamental statement that is assumed to be true without proof. For example, the statement "Two distinct points are contained by one and only one line" is a postulate of Euclidian geometry.

POTENTIAL FUNCTION If a vector field $\mathbf{f}(x, y, z)$ is the gradient of a scalar function $U(x, y, z)$, then U is said to be the potential function for the field $\mathbf{f}$. For example, if U represents potential energy, then the force acting on a body is the negative of the gradient of U. A potential function for $\mathbf{f}$ can be found only if the curl of $\mathbf{f}$ is zero. Another way of stating the condition is that the line integral of $\mathbf{f}$ around a closed path must be zero.

POWER A power of a number indicates repeated multiplication. For example, "a to the third power" means "a multiplied by itself three times" ($a^3 = a \times a \times a$). Powers are written with little raised numbers known as exponents. (See **exponent**.)

POWER SERIES A series of the form

$$c_0 + c_1 x + c_2 x^2 + c_3 x^3 + \ldots$$

where the c's are constants, is said to be a power series in x.

PRECEDENCE The rules of precedence determine the order in which operations are performed in an expression. For example, in ordinary algebraic notation and many computer programming languages, exponentiations are done first; then multiplications and divisions; and finally additions and subtractions. For example, in the expression $3 + 4 \times 5^2$ the exponentiation is done first, giving the result $3 + 4 \times 25$. Then the multiplication is done, resulting in $3+100$. Finally the addition is performed, yielding the final result, 103.

An operation enclosed in parentheses is always performed before an operation that is outside the parentheses. For example, in the expression

$$3 \times (4 + 5),$$

the addition is done first, giving 3×9. Then the multiplication is performed, yielding the final result, 27.

PREMISE A premise is one of the sentences in an argument: the conclusion of the argument follows as a result of the premises. (See **logic**.)

PRIME FACTORS Any composite number can be expressed as the product of two or more prime numbers, which are called the prime factors of that number. Here are some examples of prime factors:

4	=	2×2	16	=	$2 \times 2 \times 2 \times 2$
6	=	2×3	18	=	$2 \times 3 \times 3$
8	=	$2 \times 2 \times 2$	27	=	$3 \times 3 \times 3$
9	=	3×3	32	=	$2 \times 2 \times 2 \times 2 \times 2$
10	=	5×2	48	=	$2 \times 2 \times 2 \times 2 \times 3$
12	=	$2 \times 2 \times 3$	60	=	$2 \times 2 \times 3 \times 5$
14	=	2×7	72	=	$2 \times 2 \times 2 \times 3 \times 3$
15	=	5×3	75	=	$3 \times 5 \times 5$

Figure 93 Prism

PRIME NUMBER A prime number is a natural number that has no integer factors other than itself and 1. The smallest prime numbers are 2, 3, 5, 7, 11, 13, 17, 19, 23, 29, 31, 37, 41. (See **Eratosthenes sieve**.)

PRINCIPAL VALUES The principal values of the arcsin and arctan functions lie between $-\pi/2$ and $\pi/2$. The principal values of the arccos function are between 0 and π. (See **inverse trigonometric functions**.)

PRISM A prism is a solid that is formed by the union of all the line segments that connect corresponding points on two congruent polygons that are located in parallel planes. The regions enclosed by the polygons are called the *bases*. A line segment that connects two corresponding vertices of the polygons is called a *lateral edge*. If the lateral edges are perpendicular to the planes containing the bases, then the prism is a right prism. The distance between the planes containing the bases is called the *altitude*. The volume of the prism is (base area)×(altitude).

Prisms can be classified by the shape of their bases. A prism with triangular base is a triangular prism. (See figure 93.) A cube is an example of a right square prism. Triangular prisms made of glass have an important application. If sunlight is passed through the prism, it is split up into all the colors of the rainbow (because light of different wavelengths is refracted in different amounts by the glass). (See **Snell's law; optics.**)

PROBABILITY Probability is the study of chance occurrences. Intuitively, we know that an event with a 50 percent probability is equally likely to occur or not occur. Mathematically, probability is defined in terms of a probability space, called Ω (omega), which is the set of all possible outcomes of an experiment. Let s be the number of outcomes. For example, if you flip three coins, Ω contains eight outcomes: {(HHH), (HHT), (HTH), (HTT), (THH), (THT), (TTH), (TTT)}, where H stands for heads and T stands for tails. An *event* is a subset of Ω. For example, if A is the event that two heads appear, then $A = \{(HHT), (HTH), (THH)\}$. Let $N(A)$ be the number of outcomes in A. Then the probability that the event A will occur (written as $\Pr(A)$) is defined as $\Pr(A) = N(A)/s$. In this case $N(A) = 3$ and $s = 8$, so the probability that two heads will appear if you flip three coins is $3/8$.

An important part of probability involves counting the number of possible outcomes in the probability space. (See **combinations; permutations; sampling.**)

For information on other important probabili-

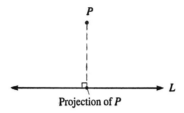

Projection of P

Figure 94

ty tools, see **random variable; discrete random variable; continuous random variable.**

Also, probability provides the foundation for **statistical inference.**

PRODUCT The product is the result obtained when two numbers are multiplied. In the equation $4 \times 5 = 20$, the number 20 is the product of 4 and 5.

PROJECTION The projection of a point P on a line L is the point on L that is cut by the line that passes through P and is perpendicular to L. See figure 94. In other words, the projection of point P is the point on line L that is the closest to point P. The projection of a set of points is the set of projections of all these points. Some shadows are examples of projections. Vectors can be projected on other vectors. (See **dot product.**)

PROLATE SPHEROID A prolate spheroid is elongaged vertically. For contrast, see **oblate spheroid.**

PROOF A proof is a sequence of statements that show a particular theorem to be true. In the course of a proof it is permissible to use only axioms (postulates) or theorems that have been previously proved.

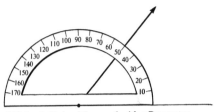

50° Angle Being Measured with a Protractor

Figure 95

PROPER FRACTION A proper fraction is a fraction with a numerator that is smaller than the denominator, for example, $\frac{2}{3}$. For contrast, see **improper fraction.**

PROPORTION A fractional equation of the form $a/b = c/d$ is called a proportion.

PROPORTIONAL If $x = ky$, where k is a constant, then x is said to be proportional to y. (See also **inversely proportional.**)

PROPOSITION A proposition is a proposed theorem that has yet to be proved.

PROTRACTOR A protractor is a device for measuring the size of angles. Put the point marked with a dot (see figure 95) at the vertex of the angle, and place the side of the protractor even with one side of the angle. Then the size of the angle can be read on the scale at the place where the other side of the angle crosses the protractor.

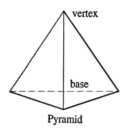

Figure 96   Pyramid

PYRAMID A pyramid (see figure 96) is formed by the union of all line segments that connect a given point (called the *vertex*) and points that lie on a given polygon. (The vertex must not be in the same plane as the polygon.) The region enclosed by the polygon is called the *base*, and the distance from the vertex to the plane containing the base is called the *altitude*. The volume of a pyramid is given by

$$(volume) = \frac{1}{3} \times (basearea) \times (altitude)$$

Pyramids are classified by the number of sides on their bases. (Note that all the faces other than the base are triangles.) A triangular pyramid, which contains four faces, is also known as a tetrahedron.

The most famous pyramids are the pyramids in Egypt. The largest of these pyramids originally had a base 756 feet square and an altitude of 481 feet.

PYTHAGORAS Pythagoras (c 580 BC to c 500 BC) was a Greek philosopher and mathematician who founded a brotherhood that developed religious and mathematical ideas. (See **Pythagorean theorem.**)

PYTHAGOREAN THEOREM The Pythagorean
theorem relates the three sides of a right triangle:

$$c^2 = a^2 + b^2$$

where c is the side opposite the right angle (the hypotenuse), and a and b are the sides adjacent to the
right angle.

For example, if the length of one leg of a right triangle is 6 and the other leg is 8, then the hypotenuse
has length $\sqrt{6^2 + 8^2} = 10$. The White House, the
Washington Monument, and the Capitol in Washington, D.C.. form a right triangle. The White House is
0.54 mile from the Washington Monument, and the
Capitol is 1.4 miles from the monument. From this
information we can determine that the distance from
the White House to the Capitol is

$$\sqrt{(0.54)^2 + (1.4)^2} = 1.5 \text{ miles.}$$

Another application of the theorem is the distance formula, which says that the distance between
two points in a plane, (x_1, y_1) and (x_2, y_2), is given
by

$$(distance) = \sqrt{(x_1 - x_2)^2 + (y_1 - y_2)^2}$$

There are many ways to prove the theorem. One
way involves similar triangles. In figure 97, triangles
ABC and ACD have exactly the same angles, so
they are similar. Since corresponding sides of similar
triangles are in proportion, we know that $c/b = b/c_2$.
Likewise, triangles ABC and CBD are similar, so
$c/a = a/c_1$. Therefore:

$$a^2 = cc_1 \text{ and } b^2 = cc_2$$

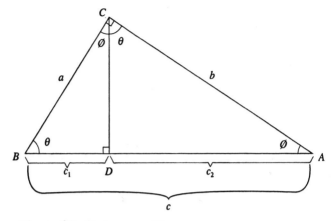

Figure 97 Pythagorean Theorem

Add these together:

$$a^2 + b^2 = c(c_1 + c_2) = c^2$$

and the theorem is demonstrated.

PYTHAGOREAN TRIPLE If three natural numbers a, b, and c satisfy $a^2 + b^2 = c^2$, then these three numbers are called a Pythagorean triple. For example, 3, 4, 5 and 5, 12, 13 are both Pythagorean triples, because $3^2 + 4^2 = 5^2$ and $5^2 + 12^2 = 13^2$.

Q

QED QED is an abbreviation for *quod erat demonstrandum*, latin for "which was to be shown." It is put at the end of a proof to signify that the proof has been completed.

QUADRANT The x- and y-axes divide a plane into four regions, each of which is called a quadrant. The four quadrants are labeled the first quadrant, the second quadrant, and so on, as shown in figure 98.

QUADRANTAL ANGLE The angles that measure $0°, 90°, 180°,$ and $270°$, and all angles coterminal with these, are called quadrantal angles.

QUADRATIC EQUATION A quadratic equation is an equation involving the second power, but no higher power, of an unknown. The general form is

$$ax^2 + bx + c = 0$$

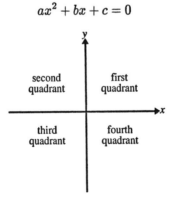

Figure 98

(a, b, and c are known; x is unknown; $a \neq 0$).

There are three ways to solve this kind of equation for x. One method is to factor the left-hand side into two linear factors. For example, to solve the equation $x^2 - 7x + 12 = 0$ we need to think of two numbers that multiply to give 12 and add to give -7. The two numbers that work are -4 and -3, which means that

$$x^2 - 7x + 12 = (x - 4)(x - 3) = 0$$

so $x = 4$ or $x = 3$.

Often the factors are too complicated to determine easily, so we need another method. One possibility is completing the square. We write the equation like this:

$$x^2 + \frac{bx}{a} = -\frac{c}{a}$$

We can simplify the equation if we can add something to the left-hand side to make it a perfect square. Add $b^2/4a^2$ to both sides:

$$x^2 + \frac{bx}{a} + \frac{b^2}{4a^2} = \frac{b^2}{4a^2} - \frac{c}{a}$$

The equation can now be rewritten as:

$$\left(x + \frac{b}{2a}\right)^2 = \frac{b^2 - 4ac}{4a^2}$$

$$x + \frac{b}{2a} = \pm\sqrt{\frac{b^2 - 4ac}{4a^2}}$$

$$x = \frac{-b \pm \sqrt{b^2 - 4ac}}{2a}$$

The last equation is known as the quadratic formula. It allows us to solve for x, given any values for a, b, and c. The third way to solve a quadratic equation is simply to remember this formula.

The formula also reveals some properties of the solutions. The key quantity is $b^2 - 4ac$, which is known as the *discriminant*. If $b^2 - 4ac$ is positive, there will be two real values for x. If $b^2 - 4ac$ has a rational square root, then x will have two rational values; otherwise x will have two irrational values. If $b^2 - 4ac$ is zero, then x will have one real value. If $b^2 - 4ac$ is negative, then x will have two complex solutions. (See **complex number.**)

The real solutions to a quadratic equation can be illustrated on a graph of Cartesian coordinates. The graph of $ax^2 + bx + c$ is a parabola. The real solutions for x will occur at the places where the parabola intersects the x-axis. Three possibilities are shown in figure 99.

QUADRATIC EQUATION, 2 UNKNOWNS

The general form of a quadratic equation in two unknowns is

(1) $Ax^2 + Bxy + Cy^2 + Dx + Ey + F = 0$

where at least one of A, B, and C is nonzero. The graph of this equation will be one of the conic sections. To determine which one, we need to write the equation in a transformed set of coordinates so we can identify the standard form of the equation. First, rotate the coordinate axes by an angle θ, where

$$\tan 2\theta = \frac{B}{A - C}$$

This procedure will get rid of the cross term Bxy. (See **rotation.**) In the new coordinates, x' and y', the equation becomes

(2) $A'x'^2 + C'y'^2 + D'x' + E'y' + F' = 0$

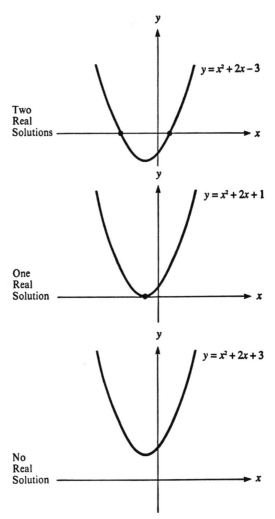

Figure 99 Quadratic equation

If either A' or C' is zero, then the graph of this equation will be a parabola. For example, suppose that there is no y'^2 term, so $C' = 0$. If you perform this translation of coordinates:

$$x'' = x' + \frac{D'}{2A'} \text{ and } y'' = y' + \frac{4A'F' - D'^2}{4A'E'}$$

the equation becomes

$$A'x''^2 + E'y'' = 0$$

which can be graphed as a parabola.

If neither A' nor C' is zero in equation (2), then perform the translation

$$x'' = x' + \frac{D'}{2A'} \text{ and } y'' = y' + \frac{E'}{2C'}$$

Then the equation can be written in the form

$$A'x''^2 + C'y''^2 + F'' = 0$$

If $A' = C'$, then this is the equation of a circle. If A' and C' have the same sign (i.e., they are both positive or both negative), the equation will be the equation of an ellipse. If A' and C' have opposite signs, the equation will be the equation of a hyperbola.

You can tell immediately what the graph of equation (1) will look like by examining the quantity $B^2 - 4AC$. It turns out that this quantity is invariant when you rotate the coordinate system. This means that $B'^2 - 4A'C'$ in equation (2) will equal $B^2 - 4AC$ in equation (1). If $B^2 - 4AC < 0$, the graph is an ellipse or a circle. If $B^2 - 4AC = 0$, the graph is a parabola. If $B^2 - 4AC > 0$, the graph is a hyperbola.

It is also possible for the solution to equation (1) to be either a pair of intersecting lines or a single point, or even for there to be no solution at all. In

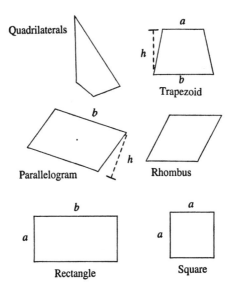

Figure 100   Quadrilaterals

these cases the solution is said to be a degenerate conic section.

QUADRATIC FORMULA The quadratic formula says that the solution for x in the equation $ax^2 + bx + c = 0$ is

$$x = \frac{-b \pm \sqrt{b^2 - 4ac}}{2a}$$

(See **quadratic equation**.)

QUADRILATERAL A quadrilateral (see figure 100) is a four-sided polygon. A quadrilateral with two sides parallel is called a *trapezoid*, with area $h(a + b)/2$. A *parallelogram* has its opposite sides paral-

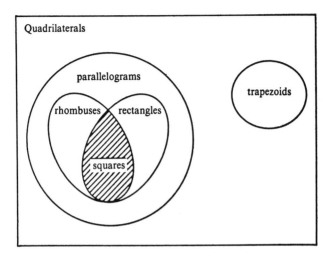

Figure 101

lel and equal. The area of a parallelogram is bh.
A quadrilateral with all four sides equal is called a
rhombus. A quadrilateral with all four angles equal
is called a *rectangle*. The sum of the four angles in a
quadrilateral is always 360°, so each angle in a rectangle is 90°. The area of a rectangle is ab. A regular
quadrilateral has all four sides and all four angles
equal, and is called a *square*. The area of a square is
a^2.

A Venn diagram (figure 101) can be used to illustrate the relationship between different types of
quadrilaterals.

QUARTIC A quartic equation is a polynomial equation of degree 4. (See **polynomial**.)

QUARTILE The first quartile of a list is the number such that one quarter of the numbers in the list are below it; the third quartile is the number such that three quarters of the numbers are below it; and the second quartile is the same as the median.

QUINTIC A quintic equation is a polynomial equation of degree 5. (See **polynomial**.)

QUOTIENT The quotient is the answer to a division problem. In the equation $33/3 = 11$, the number 11 is the quotient.

R

R^2 The R^2 value for a multiple regression is a number that indicates how well the regression explains the variance in the dependent variable. R^2 is always between 0 and 1. If it is close to 1, the regression has explained a lot of the variance; if it is close to 0, the regression has not explained very much. In the case of a simple regression, this is often written r^2, which is the square of the correlation coefficient between the independent variable and the dependent variable. (See **regression; multiple regression.**)

RADIAN MEASURE Radian measure is a way to measure angles that is often the most convenient for mathematical purposes. The radian measure of an angle is found by measuring the length of the intercepted arc and dividing it by the radius of the circle. For example, the circumference of a circle is $2\pi r$, so a full circle (360 degrees) equals 2π radians. Also, 180 degrees equals π radians, and a right angle (90 degrees) has a measure of $\pi/2$ radians. The radian measure of an angle is unit-free (i.e., it does not matter whether the radius of the circle is measured in inches, meters, or miles). Radian measure is required when trigonometric functions are used in calculus.

RADICAL The radical symbol $(\sqrt{})$ is used to indicate the taking of a root of a number. Thus $\sqrt[q]{x}$ means the qth root of x, which is the number that, when used as a factor q times, equals x: $(\sqrt[q]{x})^q = x$. Here q is called the *index* of the radical. If no index is specified, then the square root is meant. A radical always means to take the positive value. For example, both $y = 5$ and $y = -5$ satisfy $y^2 = 25$, but $\sqrt{25} = 5$. (See **root**.)

RADICAND The radicand is the part of an expression that is inside the radical sign. For example, in the expression $\sqrt{1 - x^2}$ the expression $(1 - x^2)$ is the radicand.

RADIUS The radius of a circle is the distance from the center of the circle to a point on the circle. The radius of a sphere is the distance from the center of the sphere to a point on the sphere. A line segment drawn from the center of a circle to any point on the circumference is also called a radius. The plural of "radius" is "radii."

RANDOM VARIABLE A random variable is a variable that takes on a particular value when a specified random event occurs. For example, if you flip a coin three times and X is the number of heads you toss, then X is a random variable with the possible values 0, 1, 2, and 3. In this case $\Pr(X = 0) = 1/8$, $\Pr(X = 1) = 3/8$, $\Pr(X = 2) = 3/8$, and $\Pr(X = 3) = 1/8$.

If a random variable has only a discrete number of possible values, it is called a **discrete random variable**. The probability function, or density function, for a discrete random variable is a function such that, for each possible value x_i, the value of the function is $f(x_i) = \Pr(X = x_i)$. In the three coin example, $f(0) = 1/8$, $f(1) = 3/8$, $f(2) = 3/8$, and $f(3) = 1/8$.

For examples of discrete random variable distributions, see **binomial distribution; Poisson distribution; geometric distribution; hypergeometric distribution**.

A continuous random variable is a random variable that can have many possible values over a continuous range. The density function of a continuous random variable is a function such that the area under the curve between two values gives the probability of being between those two values. (See **continuous random variable**.)

For some examples of distributions for continuous random variables, see **normal distribution; chi-square distribution; *t*-distribution; *F*-distribution**.

RANGE (1) The range of a function is the set of all possible values for the output of the function. (See **function**.)

(2) The range of a list of numbers is equal to the largest value minus the smallest value. It is a measure of the dispersion of the list—in other words, how spread out the list is.

RANK The rank of a matrix is the number of linearly independent rows it contains. The $m \times m$ matrix **A** will have rank m if all of its rows are linearly independent, as will be the case if det $\mathbf{A} \neq 0$. (See **linearly independent; determinant**.) The number of linearly independent columns in a matrix is the same as the number of linearly independent rows.

RATIO The ratio of two real numbers a and b is $a \div b$, or a/b. The ratio of a to b is sometimes written as $a : b$. For example, the ratio of the number of sides in a hexagon to the number of sides in a triangle is 6:3, which is equal to 2:1.

RATIONAL NUMBER A rational number is a number that can be expressed as the ratio of two integers. A rational number can be written in the form p/q, where p and q are both integers $(q \neq 0)$. A rational number can be expressed either as a fraction, such as $\frac{1}{5}$, or as a decimal number, such as 0.2. A fraction written in decimal form will be either a terminating decimal, such as $\frac{5}{8} = 0.625$ or $\frac{1}{4} = 0.25$, or a decimal that endlessly repeats a particular pattern, such as $\frac{1}{3} = 0.333333\ldots$, $\frac{2}{9} = 0.222222222\ldots$, $\frac{10}{11} = 0.909090909090\ldots$, or $\frac{1}{7} = 0.142857142857142857\ldots$. If the decimal representation of a number goes on forever without repeating any pattern, then that number is an **irrational number.**

RATIONAL ROOT THEOREM The rational root theorem says that, if the polynomial equation

$$a_n x^n + a_{n-1} x^{n-1} + a_{n-2} x^{n-2} + \cdots + a_2 x^2 + a_1 x + a_0 = 0$$

where $a_0, a_1, \ldots a_n$, are all integers, has any rational roots, then each rational root can be expressed as a fraction in which the numerator is a factor of a_0 and the denominator is a factor of a_n. This theorem sometimes makes it easier to find the roots of complicated polynomial equations, but it provides no help if there are no rational roots to begin with. For example, suppose that we are looking for the rational roots of the equation

$$x^3 - 9x^2 + 26x - 24 = 0$$

In this case $a_n = 1$ and $a_0 = 24$. Therefore the rational roots, if any, must have a factor of 24 in the numerator and 1 in the denominator. The factors of 24 are 1, 2, 3, 4, 6, 8, 12, 24. If we test all the possibilities, it turns out that the three roots are 2, 3, and 4.

To show that this rule holds in the case where $a_n = 1$, and all the roots are integers, note that in factored form the polynomial is:

$$(x - r_1)(x - r_2)(x - r_3) \times \cdots \times (x - r_n)$$

where r_1 to r_n are the roots of the polynomial equation. If you multiply this out, you will see that the last term becomes the product of all the roots; therefore, the roots will all be factors of a_0.

RATIONALIZING THE DENOMINATOR The process of rationalizing the denominator involves rewriting a fraction in an equivalent form that does not have an irrational number in the denominator. For example, the fraction $1/\sqrt{2}$ can be rationalized by multiplying the numerator and denominator by $\sqrt{2}$: $1/\sqrt{2} = \sqrt{2}/2$.

The fraction $1/(a + \sqrt{b})$ can be rationalized by multiplying the numerator and the denominator of the fraction by $a - \sqrt{b}$:

$$\frac{1}{a + \sqrt{b}} \times \frac{a - \sqrt{b}}{a - \sqrt{b}} = \frac{a - \sqrt{b}}{a^2 + a\sqrt{b} - a\sqrt{b} - b} = \frac{a - \sqrt{b}}{a^2 - b}$$

RAY A ray is like half of a line: it has one endpoint, and then goes off forever in a straight line. You can think of a light ray from a star as being a ray, with the endpoint located at the star.

REAL NUMBERS The set of real numbers is the set of all numbers that can be represented by points on a number line. (See figure 102.)

Figure 102 Number line for real numbers

The set of real numbers includes all rational numbers and all irrational numbers. Any real number can be expressed as a decimal fraction, which will either terminate or endlessly repeat a pattern (if the number is rational), or continue endlessly with no pattern (if the number is irrational).

Whenever the term number is used by itself, it is often assumed that the real numbers are meant. The measurement of a physical quantity, such as length, time, or energy, will be a real number.

The set of real numbers is a subset of the set of complex numbers, which includes the pure imaginary numbers plus combinations of real numbers and imaginary numbers.

RECIPROCAL The reciprocal of a number a is equal to $1/a$ (provided $a \neq 0$). For example, the reciprocal of 2 is $\frac{1}{2}$; the reciprocal of 0.01 is 100, and the reciprocal of 1 is 1. The reciprocal is the same as the multiplicative inverse.

RECTANGLE A rectangle is a quadrilateral with four 90° angles. The opposite sides of a rectangle are parallel, so the set of rectangles is a subset of the set of parallelograms. A square has four 90° angles, so the set of squares is a subset of the set of rectangles. The area of a rectangle is the product of the lengths of any two adjacent sides. For picture, see **quadrilateral**.

RECTANGULAR COORDINATES See **Cartesian coordinates.**

RECURSION Recursion is the term for a definition that refers to the object being defined. The use of a recursive definition requires care to make sure that an endless loop is not created. Here is an example of a recursive definition for the factorial function n!:

$$n! = n(n-1)!$$

(In words: "The factorial of n equals n times the factorial of $n-1$.") This definition leads to an endless loop. Here is a better recursive definition that avoids the endless loop problem:

$$\text{If } n \; > \; 0, \text{then } n! = n(n-1)!$$
$$\text{If } n \; = \; 0, \text{then } n! = 1$$

Here are the steps to use this definition to find 3!:

$$3! = 3 \times 2!$$
$$\text{Look up } 2!$$
$$2! = 2 \times 1!$$
$$\text{Look up } 1!$$
$$1! = 1 \times 0$$
$$\text{Look up } 0!$$
$$0! = 1$$
$$\text{Then } 1! = 1 \times 1$$
$$\text{Then } 2! = 2 \times 1 = 2$$
$$\text{Then } 3! = 3 \times 2 = 6$$

For an example of a geometric figure that is defined using recursion, see **fractal.**

REFLECTION A reflection is a transformation in which the transformed figure is the mirror image of

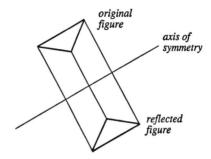

Figure 103 Reflection

the original figure. The reflection is centered on a
line called the axis of symmetry. Here is how to find
the reflection of a particular point. Draw from the
point to the axis, the line perpendicular to the axis.
Then the point on that line that is the same distance
from the axis as the original point, but on the opposite side of the axis, is the reflection of the original
point. In other words, the axis of symmetry is the
perpendicular bisector of the line segment joining a
point and its reflection. (See figure 103.)

REFLEXIVE PROPERTY The reflexive property
of equality is an axiom that states an obvious but
useful fact: $x = x$, for all x. That means that any
number is equal to itself.

REGRESSION Regression is a statistical technique
for determining the relationship between quantities.
In simple regression, there is one independent variable that is assumed to have an effect on one other variable (the dependent variable). It is necessary
to have several observations, with each observation

containing a pair of values (one for each of the two variables). The observations can be plotted on a two-dimensional diagram (see figure 104) where the independent variable, x, is measured along the horizontal axis and the dependent variable, y, is measured along the vertical axis.

The regression procedure determines the line that best fits the observations. The best-fit line is the line such that the sum of the squares of the deviations of all of the points from the line is at its minimum. The slope (m) of the best-fit line is given by the equation

$$m = \frac{\overline{xy} - \overline{x} \cdot \overline{y}}{\overline{x^2} - \overline{x}^2}$$

A bar over a quantity represents the average value of the quantity. After the slope has been found, the vertical intercept (b) of the line can be determined from the equation

$$b = \overline{y} - m\overline{x}$$

The r^2 value for the regression is a number between 0 and 1 that indicates how well the line summarizes the pattern of the observations. In some cases the line will fit the data points very well, and then the r^2 value will be close to 1. In other cases the data points cannot be well summarized by a line, and the r^2 value will be close to 0. The value of r^2 can be found from the formula:

$$r^2 = \frac{(\overline{xy} - \overline{x} \cdot \overline{y})^2}{(\overline{x^2} - \overline{x}^2)(\overline{y^2} - \overline{y}^2)}$$

Also, the r^2 value is the square of the **correlation coefficient**.

For situations where there are several independent variables, each having an effect on the dependent variable, see **multiple regression**.

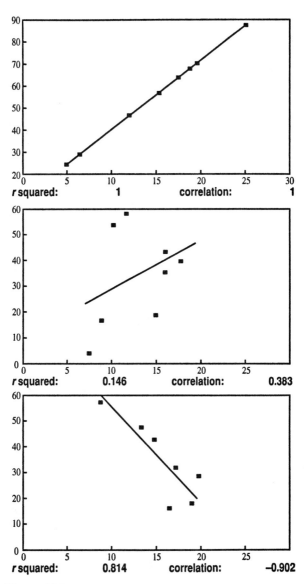

Figure 104   Regression

REGULAR POLYGON A regular polygon is a polygon in which all the angles and all the sides are equal. For example, a regular triangle is an equilateral triangle with three 60° angles. A regular quadrilateral is a square. A regular hexagon has six 120° angles.

REGULAR POLYHEDRON A regular polyhedron is a polyhedron where all faces are congruent regular polygons. There are only five possible types. (See **polyhedron.**)

REJECTION REGION The rejection region consists of those values of the test statistic for which the null hypothesis will be rejected. This is also called the critical region. (See **hypothesis testing.**)

RELATION A relation is a set of ordered pairs. The first entry in the ordered pair can be called x, and the second entry can be called y. For example, $\{(1, 0), (1, 1), (1, -1), (-1, 0)\}$ is an example of a relation. A function is also an example of a relation. A function has the special property that, for each value of x, there is a unique value of y. This property does not have to hold true for a relation. The equation of a circle $x^2 + y^2 = r^2$ defines a relation between x and y, but this relation is not a function because for every value of x there are two values of $y : \sqrt{r^2 - x^2}$ and $-\sqrt{r^2 - x^2}$.

REMAINDER In the division problem $9 \div 4$, the quotient is 2 with a remainder of 1. In general, if $m = nq + r$ (where m, n, q, and r are natural numbers and $r < n$), then the division problem m/n has the quotient q and the remainder r.

Figure 105   Rhombus

REPEATING DECIMAL A repeating decimal is
a decimal fraction in which the digits endlessly re-
peat a pattern, such as $\frac{2}{9} = 0.2222222\ldots$ or $\frac{2}{7} =$
$0.285714285714285714\ldots$. For contrast, see **terminating decimal.**

RESULTANT The resultant is the vector that results
from the addition of two other vectors.

REVERSE POLISH NOTATION Reverse Polish notation is the same as **Polish notation,** except written in reverse order: operators come after
operands.

RHOMBUS A rhombus is a quadrilateral with four
equal sides. A square is one example of a rhombus,
but in general a rhombus will look like a square that
has been bent out of shape. (See figure 105.)

RIEMANN Georg Friedrich Bernhard Riemann
(1826 to 1866) was a German mathematician who developed a version of non-Euclidian geometry in which
there are no parallel lines. This concept was used by
Einstein in the development of relativity theory. He
also made many other contributions in number theory and analysis.

RIGHT ANGLE A right angle is an angle that measures 90° ($\pi/2$ radians). It is the type of angle that makes up a square corner. (See **angle.**)

RIGHT CIRCULAR CONE A right circular cone is a cone whose base is a circle located so that the line connecting the center of the circle to the vertex of the cone is perpendicular to the plane containing the circle. (See **cone.**)

RIGHT CIRCULAR CYLINDER A right circular cylinder is a cylinder whose bases are circles and whose axis is perpendicular to the planes containing the two bases. (See **cylinder.**)

RIGHT TRIANGLE A right triangle is a triangle that contains one right angle. The side opposite the right angle is called the *hypotenuse*; the other two sides are called the legs. Since the sum of the three angles of a triangle is 180°, no triangle can contain more than one right angle. The Pythagorean theorem expresses a relationship between the three sides of a right triangle:

$$c^2 = a^2 + b^2$$

where a and b are the lengths of the two legs, and c is the length of the hypotenuse.

ROOT

(1) The root of an equation is the same as a solution to that equation. For example, the statement that a quadratic equation has two roots means that it has two solutions.

(2) The process of taking a root of a number is the opposite of raising the number to a power. The

square root of a number x (written as $\sqrt{x}$) is the number that, when raised to the second power, gives x:

$$(\sqrt{x})^2 = x$$

The symbol $\sqrt{}$ is called the *radical symbol*. A positive number has two square roots (one positive and one negative), but the radical symbol always means to take the positive square root.

Some examples of square roots are:

$$\sqrt{1} = 1,\ \sqrt{4} = 2,\ \sqrt{9} = 3,$$
$$\sqrt{16} = 4,\ \sqrt{25} = 5,\ \sqrt{36} = 6$$

A small number in front of the radical (called the index of the radical) is used to indicate that a root other than the square root is to be taken. For example $\sqrt[3]{x}$ is the cube root of x, defined so that $(\sqrt[3]{x})^3 = x$. Examples of other roots are:

$$\sqrt[3]{8} = 2,\ \sqrt[3]{27} = 3,\ \sqrt[5]{32} = 2,\ \sqrt[4]{10,000} = 10$$

Roots can also be expressed as fractional exponents:

$$\sqrt[q]{x} = x^{1/q}$$

(See **exponent**.)

ROTATION A rotation of a Cartesian coordinate system occurs when the orientation of the axes is changed but the origin is kept fixed. In figure 106 the coordinate axes x' and y' (x-prime and y-prime) are formed by rotating the original axes, x and y, by an angle θ. The main reason for doing this is that sometimes the equation for a particular figure will be much simpler in the new coordinate system than it was in the old one.

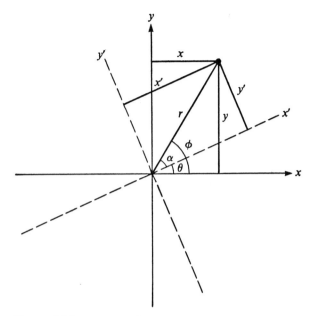

Figure 106 Rotation of coordinate axes

We need to find an expression for the new coordinates in terms of the old coordinates. Let α and ϕ be as shown in figure 106. Then $\alpha = \phi - \theta$.

From the definition of the trigonometric functions:

$$y' = r \sin \alpha, \; x' = r \cos \alpha$$

Using the formula for the sine and cosine of a difference:

$$\sin \alpha = \sin \phi \cos \theta - \cos \phi \sin \theta$$
$$\cos \alpha = \cos \phi \cos \theta + \sin \phi \sin \theta$$

Substituting:

$$y' = r \sin \phi \cos \theta - r \cos \phi \sin \theta$$

$$x' = r \cos \phi \cos \theta + r \sin \phi \sin \theta$$

Since $y = r \sin \phi$ and $x = r \cos \phi$, we can write

$$y' = y \cos \theta - x \sin \theta$$

$$x' = x \cos \theta + y \sin \theta$$

The last two equations tell us how to transform any (x, y) pair into a new (x', y') pair. We can also derive the opposite transformation:

$$y = y' \cos \theta + x' \sin \theta$$

$$x = x' \cos \theta - y' \sin \theta$$

Coordinate rotation helps considerably when we try to make a graph of the two-unknown quadratic equation

$$Ax^2 + B\,xy + Cy^2 + Dx + Ey + F = 0$$

The problem is caused by the $B\,xy$ term. If that term weren't present, the equation could be graphed as a conic section. Therefore what we would like to do is to choose some angle of rotation θ so that the equation written in the new coordinates will not have any $x'y'$ term. We can use the rotation transformation to find out what the equation will be in the new coordinate system:

$$
\begin{aligned}
x &= x' \cos \theta - y' \sin \theta \\
y &= y' \cos \theta + x' \sin \theta \\
xy &= x'y' \cos^2 \theta + x'^2 \sin \theta \cos \theta - \\
    &\quad y'^2 \sin \theta \cos \theta - y'x' \sin^2 \theta \\
x^2 &= x'^2 \cos^2 \theta - 2x'y' \cos \theta \sin \theta + y'^2 \sin^2 \theta \\
y^2 &= y'^2 \cos^2 \theta + 2x'y' \cos \theta \sin \theta + x'^2 \sin^2 \theta
\end{aligned}
$$

After we have combined all these terms, the equation becomes

$$x'^2[A\cos^2\theta+C\sin^2\theta+B\sin\theta\cos\theta]+x'[D\cos\theta+E\sin\theta]$$

$$+y'^2[A\sin^2\theta+C\cos^2\theta-B\sin\theta\cos\theta]+y'[-D\sin\theta+E\cos\theta]$$

$$+x'y'[-2A\cos\theta\sin\theta+2C\cos\theta\sin\theta+B\cos^2\theta-B\sin^2\theta]$$

$$+F=0$$

To get rid of the $x'y'$ term, we need to choose θ so that

$$
\begin{aligned}
0 &= 2\cos\theta\sin\theta(C-A)+B(\cos^2\theta-\sin^2\theta)\\
0 &= (C-A)\sin 2\theta+B\cos 2\theta\\
\frac{\sin 2\theta}{\cos 2\theta} &= \frac{B}{A-C}\\
\tan 2\theta &= \frac{B}{A-C}\\
\theta &= \frac{1}{2}\arctan\frac{B}{A-C}
\end{aligned}
$$

For an example of a rotation, consider the equation $xy=1$. Here, we have $B=1, F=-1$, and $A=C=D=E=0$. To choose θ so as to eliminate the cross term, we must have $\theta=\frac{1}{2}\arctan(1/0)$, or $\theta=\pi/4=45°$.

To find the equation in terms of x' and y', use the rotation transformation:

$$x=2^{-1/2}(x'-y'),\ y=2^{-1/2}(y'+x')$$

(Use the fact that $\sin\pi/4=\cos\pi/4=2^{-1/2}$.)
The rotated equation becomes:

$$1=\frac{1}{2}x'^2-\frac{1}{2}y'^2$$

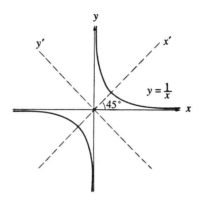

Figure 107

which is the standard form for the equation of a hyperbola. (See figure 107.)

ROUNDING Rounding provides a way of approximating a number in a form with fewer digits. A
number can be rounded to the nearest integer, or
it can be rounded to a specified number of decimal
places, or it can be rounded to the nearest number
that is a multiple of a power of 10. For example, 3.52
rounded to the nearest integer is 4; 6.37 rounded to
the nearest integer is 6. If 3.52 is rounded to one decimal place, the result is 3.5: if 6.37 is rounded to one
decimal place, the result is 6.4. The number 343,619
becomes 344,000 when it is rounded to the nearest
thousand. It is often helpful to present the final result of a calculation in rounded form, but the results
of intermediate calculations should not be rounded
because rounding could lead to an accumulation of
errors.

S

SADDLE POINT A saddle point is a critical point
that is not a maximum or minimum. For example, if
$f(x,y) = x^2 - y^2$, then both first partial derivatives
are zero at the point (0,0). The curve is a minimum
point if you cut a cross section along the x-axis, but it
is a maximum point if you cut a cross section along
the y-axis. Therefore, it is a saddle point. To see
where the name comes from, imagine you are an ant
in the middle of a saddle on a horse's back. If you
look toward the front or back of the horse, you will
seem to be in the bottom of a valley—that is, a minimum point. However, if you look in the direction of
the sides of the horse, you will seem to be at the top
of a hill—a maximum. (See **second-order conditions**.)

SAMPLE A sample is a group of items chosen from
a population. The characteristics of the sample are
used to estimate the characteristics of the population. (See **sampling; statistical inference**.)

SAMPLING To sample j items from a population of
n objects with replacement means to choose an item,
then replace the item, and repeat the process j times.
Flipping a coin 1 time is equivalent to sampling with
replacement from a population of size 2. The fact
that you've flipped heads once does not mean that
you cannot flip heads the next time. There are n^j
possible ways of selecting a sample of size j from a
population of size n with replacement.

To sample j items from a population of n objects

without replacement means to select an item, and
then select another item from the remaining $n-1$ objects, and repeat the process j times. Dealing a poker
hand is an example of sampling without replacement
from a population of 52 objects. After you've dealt
the first card, you can't deal that card again, so there
are 51 possibilities for the second card. There are
$n!/(n-j)!$ ways of selecting j items from a population of size n without replacement.

The concept of the two different kinds of sampling provides the answer to the birthday problem
in probability. Suppose that you have a group of s
people. What is the probability that no two people in
the group will have the same birthday? The number
of possible ways of distributing the birthdays among
the s people is 365^s. (That is the same as sampling
s times from a population of size 365 with replacement.) To find the number of ways of distributing
the birthdays so that nobody has the same birthday,
you have to find out how many ways there are of
sampling s items from a population of 365 without
replacement, which is $365!/(365-s)!$. The probability that no two people will have the same birthday
is therefore

$$\frac{365!/(365-s)!}{365^s}$$

For example, if $s=3$, the formula gives the probability

$$\frac{365 \times 364 \times 363}{365 \times 365 \times 365} = .992$$

The table gives the value of this probability for
different values of s.

s	Probability
2	.997
3	.992
5	.973
10	.883
15	.747
20	.589
30	.294
50	.030

This result says that in a group of 50 people there is only a 3 percent chance that they will all have different birthdays.

(See also **combinations; permutations**.)

SCALAR A scalar is a quantity that has size but not direction. For example, real numbers are scalars. By contrast, a vector has both size and direction.

SCALAR PRODUCT The scalar product (or dot product) of two vectors (x_1, y_1, z_1) and (x_2, y_2, z_2) is defined to be $(x_1 x_2 + y_1 y_2 + z_1 z_2)$ This quantity is a number (a scalar) rather than a vector. (See **dot product**.)

SCALENE TRIANGLE A scalene triangle is a triangle in which no two sides are equal.

SCIENTIFIC NOTATION Scientific notation is a short-hand way of writing very large or very small numbers. A number expressed in scientific notation is expressed as a number between 1 and 10 multiplied by a power of 10. For example, the number of meters in a light year is about 9,460,000,000,000,000. It is much easier to write this number as 9.46×10^{15}. The

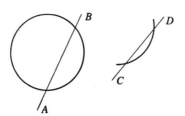

Figure 108 Secant lines

wavelength of red light is 0.0000007 meters, which can be written in scientific notation as 7×10^{-7} meter. Computers use a form of scientific notation for big numbers, as do some pocket calculators.

SECANT (1) A secant line is a line that intersects a circle, or some other curve, in two places. Lines *AB* and *CD* in figure 108 are both secant lines. For contrast, see **tangent**.

(2) The secant function is defined as the reciprocal of the cosine function: $\sec\theta = 1/\cos\theta$. (See **trigonometry**.)

SECOND A second is a unit of measure of an angle equal to 1/60 of a minute (or 1/3600 of a degree).

SECOND-ORDER CONDITIONS The second-order conditions are used to distinguish whether a critical point is a maximum or a minimum. To see the case of one variable, see **extremum.** With two variables, it is more complicated. Let $f(x,y)$ be a function with two variables, and suppose that both partial derivatives $(\partial f/\partial x)$ and $(\partial f/\partial y)$ are zero at a point (x_1, y_1). Use the following notation for the three second-order derivatives:

$$f_{xx} = \frac{\partial^2 f}{\partial x^2}$$

$$f_{yy} = \frac{\partial^2 f}{\partial y^2}$$

$$f_{xy} = \frac{\partial^2 f}{\partial x \partial y}$$

Evaluate each of these at the point (x_1, y_1). There are three cases to consider:

(1) If $f_{xx} f_{yy} > (f_{xy})^2$, there is a local maximum or minimum. To tell the difference: If f_{xx} and f_{yy} are positive, you have a local minimum. This means that a cross-section of the curve will be concave upward. If f_{xx} and f_{yy} are negative, you have a local maximum.

(2) If $(f_{xy})^2 > f_{xx} f_{yy}$, there is a **saddle point.**

(3) If $(f_{xy})^2 = f_{xx} f_{yy}$ you cannot tell from this test whether you have a maximum, minimum, or saddle point.

SECTOR A sector of a circle is a region bounded by two radii of the circle and by the arc of the circle whose endpoints lie on those radii. In other words, a sector is shaped like a pie slice. (See figure 109.) If r is the radius of the circle and θ is the angle between the two radii (measured in radians), then the area of the sector is $\frac{1}{2}\theta r^2$.

SEGMENT (1) The segment AB is the union of point A and point B and all points between them. (See **between.**) A segment is like a piece of a straight line. A segment has two endpoints, whereas a line goes off to infinity in two directions.

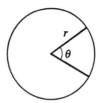

Figure 109   Sector of circle

(2) A segment of a circle is an area bounded by an arc and the chord that connects the two endpoints of the arc.

SEMILOG GRAPH PAPER Semilog graph paper has a logarithmic scale on one axis, and a uniform scale on the other axis. It is useful for graphing equations like $y = ck^x$.

SEMIMAJOR AXIS The semimajor axis of an ellipse is equal to one half of the longest distance across the ellipse.

SEMIMINOR AXIS The semiminor axis of an ellipse is equal to one half of the shortest distance across the ellipse.

SENTENCE See logic.

SEQUENCE A sequence is a set of numbers in which the numbers have a prescribed order. Some common examples of sequences are arithmetic sequences (where the difference between successive terms is constant) and geometric sequences (where the ratio between successive terms is constant). If all the terms in a sequence are to be added, it is called a **series**.

SERIES A series is the indicated sum of a sequence of numbers. Examples of series are as follows:

$$1 + 3 + 5 + 7 + 9 + 11 + 13$$
$$a + (a + b) + (a + 2b) + (a + 3b) + \ldots + [a + (n - 1)b]$$
$$2 + 4 + 8 + 16 + 32 + 64$$
$$a + ar + ar^2 + ar^3 + \ldots + ar^{n-1}$$

The first two series are examples of **arithmetic series**. The last two series are examples of **geometric series**. For other important types of series, see **Taylor series; power series**.

SET A set is a well-defined group of objects. A set is described by some rule that makes it possible to tell whether or not a particular object is in the set. The set of all natural numbers less than 11 consists of {1, 2, 3, 4, 5, 6, 7, 8, 9, 10}. Sets can be defined by listing all their elements within braces, such as {New York City, Los Angeles, Chicago}, or by giving a description that determines what is in the set and what is not: "An ellipse is the set of all points in a plane such that the sum of the distances to two fixed points in the plane is a constant." The relationship between sets can be indicated on a type of diagram known as a Venn diagram. (See figure 110.) (See also **intersection; union**.)

SEXAGESIMAL SYSTEM The basic unit in the sexagesimal system for measuring angles is the degree. If you place a one degree (1°) angle in the

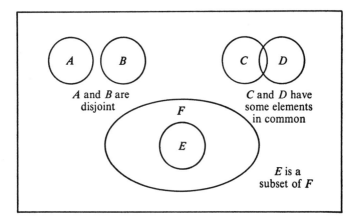

Figure 110

center of a circle, the angle will cut across 1/360 of
the circumference of the circle.

SIGMA (1) The Greek capital letter sigma (Σ) is
used to indicate summation. (See **summation notation.**)

(2) The lower case letter sigma (σ) is used to
indicate **standard deviation.**

SIGN The sign of a number is the symbol that tells
whether the number is positive ($+$) or negative ($-$).

SIGNIFICANT DIGITS The number of significant
digits expressed in a measurement indicates how pre-
cise that measurement is. A nonzero digit is always
a significant digit.

Trailing zeros to the left of the decimal point
are not significant if there are no digits to the right
of the decimal point. For example, the number

243,000,000 contains three significant digits; this means that the true value of the measurement is between 242,500,000 and 243,500,000.

Trailing zeros to the right of the decimal point are significant. For example, the number 2.1300 has five significant digits; this means that the true value is between 2.12995 and 2.13005.

Do not include more significant digits in the result of a calculation than were present in the original measurement. For example, if you calculate 243,000,000/7, do not express the result as 34,714,286, since you do not have eight significant digits to work from. Instead, express the result as 34,700,000, which, like the original measurement, has three significant digits. (However, if a calculation involves several steps, you should retain more digits during the intermediate stages.)

SIMILAR Two polygons are similar if they have exactly the same shape, but different sizes. (See figure 111.) For example, suppose you look at a color slide showing a picture of a house shaped like a rectangle. If you put the slide into a projector, you will then see on the screen a much bigger image of the same rectangle. These two rectangles are similar. Each angle in the little polygon will be equal to a corresponding angle in the big polygon. Each side on the little polygon will have a corresponding side on the big polygon. If one side of the little polygon is half as big as its corresponding side on the big polygon, then all the sides on the little polygon will be half as big as the corresponding sides on the big polygon. For polygons that have the same size, as well as the same shape, see **congruent**.

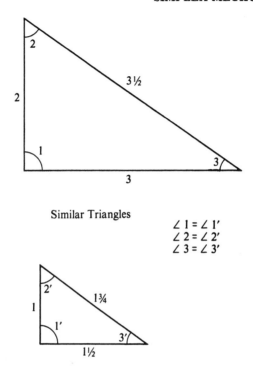

Similar Triangles

∠ 1 = ∠ 1'
∠ 2 = ∠ 2'
∠ 3 = ∠ 3'

Figure 111

SIMPLEX METHOD The simplex method, developed by mathematician George Dantzig, is a procedure for solving linear programming problems. (See **linear programming**.) The method starts by identifying a point that is one of the basic feasible solutions to the problem. (See **basic feasible solution**.) Then it provides a procedure to test whether that point is the optimal solution. If it is not, then it provides a procedure for moving to a new basic feasible solution that will have a better value for the objective function. (If you are trying to maximize the objective function, then you want to move to a point with a larger objective function value.) The procedure described above is repeated until the optimal solution has been found. In practice the calculations are usually performed by a computer.

SIMULTANEOUS EQUATIONS A system of simultaneous equations is a group of equations that must all be true at the same time. If there are more unknowns then there are equations, there will usually be many possible solutions. For example, in the two-unknown, one-equation system $x + y = 5$, there will be an infinite number of solutions, all lying along a line. If there are more equations than there are unknowns, there will often be contradictory equations, which means that no solution is possible. For example, the two-equation, one-unknown system

$$2x = 10$$

$$3x = 10$$

clearly has no solution that will satisfy both equations simultaneously. For there to be a unique solution to a system, there must be exactly as many distinct equations as there are unknowns. For example, the two-equation, two-unknown system

$$3x + 2y = 33$$

$$-x + y = 4$$

has the unique solution $x = 5, y = 9$.

When counting equations, though, you have to be careful to avoid counting equations that are redundant. For example, if you look closely at the three-equation, three-unknown system

$$2x + y + 3z \ = \ 9$$
$$4x + 9y + 0.5z \ = \ 1$$
$$2x + y + 3z \ = \ 9$$

you will see that the first equation and the last equation are exactly the same. This means that there really are only two distinct equations. Equations can be redundant even if they are not exactly the same. If one equation can be written as a multiple of another equation, then the two equations are equivalent and therefore the second equation is redundant. For example, these two equations:

$$x + y + z \ = \ 1$$
$$2x + 2y + 2z \ = \ 2$$

say exactly the same thing.

Also, if an equation can be written as a linear combination of some of the other equations in the system, then it is redundant.

A linear equation is an equation that does not have any unknowns raised to any power (other than 1). Systems of simultaneous nonlinear equations can be very difficult to solve, but there are standard ways for solving simultaneous equations if all the equations are linear.

Simple systems can be solved by the method of substitution. For example, to solve the system

$$2x + y = 9$$

$$x + 3y = 17$$

first solve the second equation for x: $x = 17 - 3y$. Now, substitute this expression for x back into the

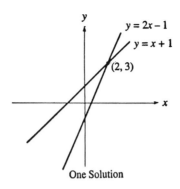

One Solution

Figure 112

first equation, and the result is a one-unknown equation: $2(17 - 3y) + y = 9$. That equation can be solved to find $y = 5$. The value of x can be found by substituting this value for y into the second equation: $x = 2$. The substitution method is often the simplest for two-equation systems, but it can be very cumbersome for longer systems.

If the simultaneous equation is written in matrix form: $\mathbf{Ax} = \mathbf{b}$, where $\mathbf{A}$ is an $n \times n$ matrix of known coefficients, $\mathbf{x}$ is an $n \times 1$ matrix of unknowns, and $\mathbf{b}$ is an $n \times 1$ matrix of known constants, then the solution can be found by finding the inverse matrix $\mathbf{A}^{-1}$:

$$\mathbf{x} = \mathbf{A}^{-1}\mathbf{b}$$

However, if the determinant of $\mathbf{A}$ is zero, then the inverse of $\mathbf{A}$ does not exist, which means either that the equations contradict each other (meaning that there is no solution), or that there is an infinite number of solutions. (See **matrix; matrix multiplication; Cramer's rule; Gauss-Jordan elimination.**)

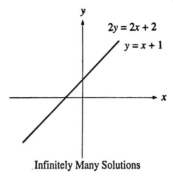

Infinitely Many Solutions

Figure 113

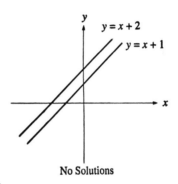

No Solutions

Figure 114

A two-equation system can also be solved by graphing. A linear equation in two unknowns defines a line. The solution to a two-equation system occurs at the point of intersection between the two lines (figure 112). If the two equations are redundant, then they define the same line, so there is an infinite number of solutions (figure 113). If the two equations are contradictory, then their graphs will be parallel lines, meaning that there will be no intersection and no solution (figure 114).

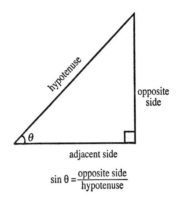

$$\sin \theta = \frac{\text{opposite side}}{\text{hypotenuse}}$$

Figure 115

SINE The sine of angle θ that occurs in a right triangle
is defined to be the length of the opposite side divided
by the length of the hypotenuse. (See figure 115.)

For a general angle in standard position (that is,
its vertex is at the origin and its initial side is along
the x axis), pick any point on the terminal side of
the angle, and then $\sin \theta = y/r$. (See figure 116.)

The table gives some special values of $\sin \theta$. (See
figure 117.)

θ (degrees)	θ (radians)	$\sin \theta$
0	0	0
30	$\pi/6$	$1/2$
45	$\pi/4$	$1/\sqrt{2}$
60	$\pi/3$	$\sqrt{3}/2$
90	$\pi/2$	1
180	π	0
270	$3\pi/2$	-1
360	2π	0

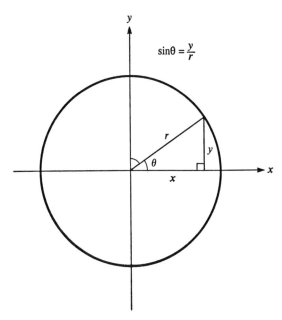

Figure 116

For most other values of θ there is no simple algebraic expression for $\sin \theta$. If θ is measured in radians, then we can find the value for $\sin \theta$ from the series

$$\sin \theta = \theta - \frac{\theta^3}{3!} + \frac{\theta^5}{5!} - \frac{\theta^7}{7!} + \cdots$$

(See **Taylor series**.)

Figure 118 shows a graph of the sine function (x is measured in radians). The value of $\sin x$ is always between -1 and 1, and the function is periodic because

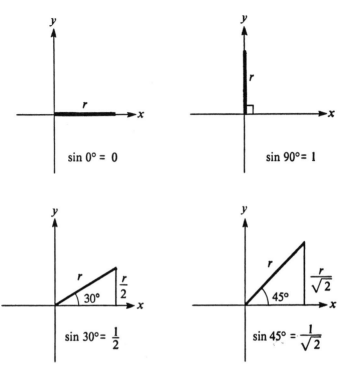

Figure 117

$$\sin x = \sin(x + 2\pi) = \sin(x + 4\pi) = \sin(x + 6\pi)$$

and so on.

Because the graph of a sine wave oscillates smoothly back and forth, the sine function describes wave patterns, harmonic motion, and voltage in alternating-current circuits.

To learn how the sine function relates to the other trigonometric functions, see **trigonometry.**

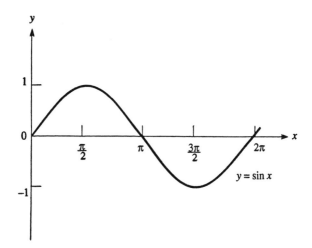

Figure 118

SKEW Two lines are skew if they are not in the same plane. Any pair of lines will either intersect, be parallel, or be skew.

SLACK VARIABLE A slack variable is a variable that is added to a linear programming problem that measures the excess capacity associated with a constraint. (See **linear programming**.)

SLIDE RULE A slide rule is a calculating device consisting of two sliding logarithmic scales. Since $\log(ab) = \log a + \log b$, a slide rule can be used to convert a multiplication problem into an addition problem, which can be performed by sliding one scale along the other. (See figure 119.) Slide rules were commonly used before pocket calculators became available.

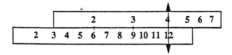

Logarithmic scale slide rule for multiplication: $3 \times 4 = 12$.

Figure 119

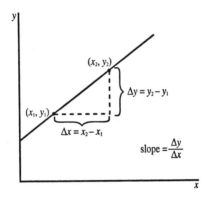

Figure 120

SLOPE The slope of a line is a number that measures how steep the line is. A horizontal line has a slope of zero. A vertical line has an infinite slope. The slope of a line is defined to be $\Delta y/\Delta x$, where Δy is the change in the vertical coordinate and Δx is the change in the horizontal coordinate between any two points on the line. (See figure 120.) The slope of the line $y = mx + b$ is m. To find the slope of a curve, see **calculus**.

SNELL'S LAW When a light ray passes from one medium to another, then it will be bent by an

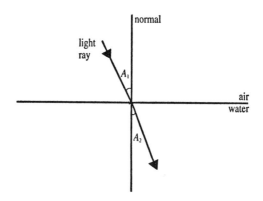

Figure 121 Snell's law

amount given by Snell's law. For every medium
through which light travels, it is possible to define
a quantity known as the index of refraction, which is
a measure of how much the speed of light is slowed
down in that medium. The index of refraction for
a pure vacuum is 1; the index of refraction of air is
very close to 1. The index of refraction of water is
1.33.

Suppose that a light ray is passing from medium
1, with index of refraction n_1, into medium 2, with
index of refraction n_2. Let A_1 by the angle of incidence (that is, the angle between the light ray and
the normal line in the first medium) and let A_2 be
the angle of refraction (the angle between the light
ray and the normal in the second medium). (See
figure 121.) Then Snell's law states that

$$n_1 \sin A_1 = n_2 \sin A_2$$

For example, if a light ray passes from air to water

at an angle of incidence of 30°, then the angle of refraction will be

$$\arcsin\left(\frac{1 \cdot \sin 30°}{1.33}\right) = \arcsin .376 = 22.1°$$

If you hold a stick in water, it will appear to be bent. (See also **optics**.)

SOLID A solid is a three-dimensional geometric figure that completely encloses a volume of space. A cereal box is an example of a solid, but a cereal bowl is not. For examples of solids, see **prism; sphere; cylinder; cone; pyramid;** and **polyhedron**.

SOLUTION If the value x_1 makes an equation involving x true, then x_1 is a solution of the equation. For example, the value 4 is a solution to the equation $x + 5 = 9$, and -3 and 3 are both solutions of the equation $x^2 - 9 = 0$. The set of all solutions to an equation is called the solution set.

If you have more than one equation with more than one unknown, see **simultaneous equations.**

SOLUTION SET The solution set for an equation consists of all of the values of the unknowns that make the equation true.

SOLVE To solve an equation means to find the solutions for the equation (i.e., to find the values of the unknowns that make the equation true).

SOLVING TRIANGLES The following rules tell how to solve for the unknown parts of a triangle:

1. If you know two angles of a triangle, you can easily find the third angle (since the sum of the three angles must be 180°).

2. If you know the three angles of a triangle but do not know the length of any of the sides, you can determine the shape of the triangle, but you have no idea about its size.

3. If you know the length of two sides (a and b) and the size of the angle between those two sides (C), then you can solve for the third side (c) by using the law of cosines:

$$c^2 = a^2 + b^2 - 2ab \cos C$$

4. If you know the length of one side (a) and the two angles next to that side (B and C), you can find the third angle ($A = 180° - B - C$), then use the law of sines to find the remaining sides:

$$b = a \sin B / \sin A$$

$$c = a \sin C / \sin A$$

5. If you know the length of the three sides, then use the law of cosines to find the cosine of the angles:

$$\cos C = \frac{a^2 + b^2 - c^2}{2ab}$$

You may find similar expressions for $\cos A$ and $\cos B$.

6. If you know the length of two sides (b and c) and the size of one angle other than the one between those two sides, there are three possibilities. Suppose you know angle B. Then use the law of sines:

$$\sin C = \frac{c \sin B}{b}$$

—If $c \sin B / b$ is less than 1, then there are two possible values for C, one obtuse and one acute, and there are two triangles that satisfy the given specifications. This is called the ambiguous case.

—If $c \sin B / b = 1$, then C is a right angle, and there is only one triangle that satisfies the given specifications.

—If $c \sin B / b$ is greater than 1, there is no triangle that satisfies the given specifications (since $\sin C$ cannot be greater than 1).

SPEED The speed of an object is the magnitude of its velocity. (See **velocity**.)

SPHERE A sphere is the set of all points in three-dimensional space that are a fixed distance from a given point (called the *center*). Some obvious examples of spheres include basketballs, baseballs, tennis balls, and (almost) the Earth. The distance from the center to any point on the sphere is called the *radius*. The distance across the sphere through the center is called the *diameter*.

The intersection between a sphere and a plane is a circle. The intersection between a sphere and a plane passing through the center is called a *great circle*. A great circle is larger than any other possible circle formed by intersecting the sphere by a plane. The shortest distance along the sphere between two points on the sphere is the path formed by the great circle that connects those two points. (See **spherical trigonometry**.)

The circumference of a great circle is also known as the circumference of the sphere. The circumference of the Earth is about 24,900 miles. The volume of a sphere is $\frac{4}{3}\pi r^3$, where r is the radius. (See **volume, figure of revolution**.) The surface area of a sphere is $4\pi r^2$. (See **surface area, figure of revolution**.)

SPHERICAL   TRIGONOMETRY   Spherical
trigonometry is the study of triangles located on the
surface of a sphere. (By contrast, ordinary trigonometry is concerned with triangles located on a plane.)
Spherical trigonometry has many applications involving navigating along the spherical surface of the
earth.

Like a plane triangle, a spherical triangle has
three vertices and three sides. However, unlike a
plane triangle, the sides are not straight lines; instead, each side is a **great circle** path connecting
two of the vertices. Since each side is an arc of a
circle, its size can be expressed in degree measure or
radian measure. There is a **dihedral angle** at each
vertex, formed by the two planes containing the great
circles representing the two sides that meet at that
vertex. It is customary to use capital letters to represent the angles in the triangle, and small letters to
represent the degree measure of the three sides. Side
a is opposite angle A, side b is opposite angle B, and
side c is opposite angle C. (See figure 122.)

The three angles in a spherical triangle add up
to more than $180°$. It is even possible to have a
spherical triangle with three right angles. (For example, consider a spherical triangle with one vertex
at the north pole, another vertex on the equator at
latitude $= 0$, longitude $= 0$, and the other vertex
along the equator at latitude $= 0$, longitude $= 90°$.)
However, a small spherical triangle will be similar to
a plane triangle, and its three angles will add up to
only slightly more than 180 degrees.

Consider a spherical right triangle, where C is the
right angle. These formulas apply:

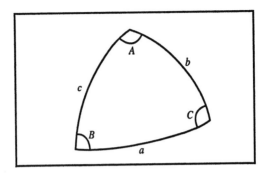

Figure 122 Spherical triangle

Spherical right triangle:

$$\cos c = \cos a \cos b$$

$$\cos c = \operatorname{ctn} A \operatorname{ctn} B$$

Formulas for angle A Formulas for angle B

$$\sin A = \frac{\sin a}{\sin c} \qquad\qquad \sin B = \frac{\sin b}{\sin c}$$

$$\cos A = \frac{\tan b}{\tan c} \qquad\qquad \cos B = \frac{\tan a}{\tan c}$$

$$\tan A = \frac{\tan a}{\sin b} \qquad\qquad \tan B = \frac{\tan b}{\sin a}$$

$$\sin A = \frac{\cos B}{\cos b} \qquad\qquad \sin B = \frac{\cos A}{\cos a}$$

The following formulas apply for all spherical triangles:

Law of Sines for Spherical Triangles

$$\frac{\sin a}{\sin A} = \frac{\sin b}{\sin B} = \frac{\sin c}{\sin C}$$

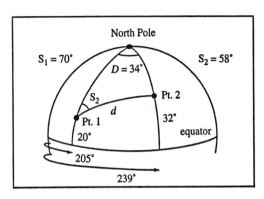

Figure 123 Spherical Triangle on Surface of Earth

Law of Cosines for Sides for Spherical Triangles:

$$\cos c = \cos a \cos b + \sin a \sin b \cos C$$

Law of Cosines for Angles for Spherical Triangles:

$$\cos C = -\cos A \cos B + \sin A \sin B \cos c$$

For example, suppose you need to calculate the shortest possible distance along the surface of the Earth between point 1 (longitude lon_1, latitude lat_1) and point 2 (coordinates lon_2, lat_2.). Set up the spherical triangle with the north pole as one vertex, and these two points as the other vertices. Then the three sides of the spherical triangle are:

$$s_1 = 90° - lat_1$$
$$s_2 = 90° - lat_2$$

d = the distance between the two points along the great circle route—that is, the result we are looking for. Angle D is the difference in longitude between the two points:

$$D = lon_2 - lon_1$$

(See figure 123.)

From the law of cosines for sides:

$$\cos d = \cos s_1 \cos s_2 + \sin s_1 \sin s_2 \cos D$$

If we measure side d in radians, the distance is rd, where r is the radius of the Earth (6375 kilometers). Then we have this formula:

distance =

$$r \arccos[(\sin lat_1 \sin lat_2) + (\cos lat_1 \cos lat_2 \cos D)]$$

For example, if point 1 is at longitude 205°, latitude 20°, and point 2 is at longitude 239° and latitude 32°, the distance between them is:

$$6375 \arccos[\sin 20° \sin 32° + \cos 20° \cos 32° \cos 34°]$$

$$= 6375 \arccos .8419 = 6375 \times .5700 = 3634 \text{ kilometers}$$

SPHEROID A spheroid is similar to a sphere but is lengthened or shortened in one dimension. (See **ellipsoid; prolate spheriod; oblate spheroid.**)

SPIRAL The curve $r = a\theta$, graphed in polar coordinates, has a spiral shape. (See figure 124.)

SQUARE (1) A square is a quadrilateral with four 90° angles and four equal sides. Chessboards are made up of 64 squares. (See **quadrilateral.**)

(2) The square of a number is found by multiplying that number by itself. For example, 4 squared

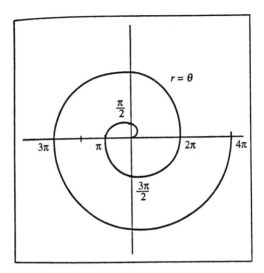

Figure 124 Spiral

equals 4 times 4, which is 16. If a square is formed
with sides a units long, then the area of that square
is a squared (written as a^2).

SQUARE MATRIX A square matrix has equal
number of rows and columns. (See **matrix; determinant.**)

SQUARE ROOT The square root of a number x
(written as $\sqrt{x}$) is the number that, when multiplied
by itself, gives x:

$$(\sqrt{x}) \times (\sqrt{x}) = (\sqrt{x})^2 = x$$

For example, $\sqrt{36} = 6$ because $6 \times 6 = 36$. Any
positive number has two square roots: one positive

and one negative. The square root symbol always means to take the positive value of the square root. (See **root**.) To find $\sqrt{x}$ when x is negative, see **imaginary number**.

The square roots of most integers are irrational numbers. For example, the square root of 2 can be approximated by $\sqrt{2} = 1.41421356\ldots$

Square roots obey the property that

$$\sqrt{ab} = \sqrt{a} \cdot \sqrt{b}$$

For example:

$$\sqrt{225} = \sqrt{9 \times 25} = \sqrt{9} \times \sqrt{25} = 3 \times 5 = 15$$

STANDARD DEVIATION The standard deviation of a random variable or list of numbers (usually symbolized by the Greek lower-case letter sigma: σ) is the square root of the variance. (See **variance**.)

The standard deviation of the list $x_1, x_2, x_3 \ldots x_n$ is given by the formula:

$$\sigma = \sqrt{\frac{(x_1 - \overline{x})^2 + (x_2 - \overline{x})^2 + \ldots + (x_n - \overline{x})^2}{n}}$$

where $\overline{x}$ is the average of the x's. The above formula is used when you know all of the values in the population. If, instead, the values $x_1 \ldots x_n$ come from a random sample chosen from the population, then the sample standard deviation is calculated, which uses the same formula as above except that $(n-1)$ is used instead of n in the denominator.

STANDARD POSITION An angle is in standard position if its vertex is at the origin and its initial side is along the x-axis.

STATISTIC A statistic is a quantity calculated from the items in a sample. For example, the average of a set of numbers is a statistic. In statistical inference, the value of a statistic is often used as an estimator of the unknown value of a population parameter.

STATISTICAL INFERENCE Statistical inference refers to the process of estimating unobservable characteristics on the basis of information that can be observed. The complete set of all items of interest is called the *population*. The characteristics of the population are usually not known. In most cases it is too expensive to survey the entire population. However, it is possible to obtain information on a group randomly selected from the population. This group is called a *sample*. For example, a pollster trying to predict the results of an election will interview a randomly selected sample of voters.

An unknown characteristic of a population is called a *parameter*. Here are two examples of parameters:

The fraction of voters in the state who support candidate X,

The mean height of all nine-year-olds in the country.

A quantity that is calculated from a sample is called a *statistic*. Here are two examples of statistics:

The fraction of voters in a 200-person poll who support candidate X,

The mean height in a randomly selected group of 90 nine-year-olds.

In many cases the value of a statistic is used as an indicator of the value of a parameter. This

type of statistic is called an *estimator*. In some cases
it is fairly obvious which estimator should be used.
For example, we would use the fraction of voters in
the sample who support candidate X as an estimator for the fraction of voters in the population supporting that candidate, and we would use the mean
height of 9-year-olds in the sample as an estimator
for the mean height of 9-year-olds in the population.
In the formal theory of statistics, certain properties
have been found to be characteristic of good esti-
mators. (See **consistent estimator; maximum
likelihood estimator; unbiased estimator.**) In
some cases, as in both of the examples given above,
an estimator has all of these desirable properties; in
other cases it is not possible to find a single estimator that has all of them. Then it is more difficult to
select the best estimator to use.

After calculating the value of an estimator, it is
also necessary to determine whether that estimator
is very reliable. If the fraction of voters in our sample who support candidate X is much different from
the fraction of voters in the population, then our estimator will give us a very misleading result. There
is no way to know with certainty whether an estimator is reliable, since the true value of the population
parameter is unknown. However, the use of statistical inference provides some indication as to the re-
liability of an estimator. First, it is very important
that the sample be selected randomly. For example,
if we select the first 200 adults that we meet on the
street, but it turns out that the street we chose is
around the the corner from candidate X's campaign
headquarters, our sample will be highly unrepresen-
tative. The best way to choose the sample would be

to list the names of everyone in the population on little balls, put the balls in a big drum, mix them very thoroughly, and then select 200 balls to represent the people in the sample. That method is not very practical, but modern pollsters use methods that are based on similar concepts of random selection.

It is important to realize that pseudo-polls, such as television call-in polls, have made no effort to make a random selection, so these are totally worthless and misleading samples.

If the sample has been selected randomly, then the methods of probability can be used to determine the likely composition of the sample. Statistical inference is based on probability. Suppose a poll found that 45 percent of the people in the sample support candidate X. If the poll is a good one, the announced result will include a statement similar to this: "There is a 95 percent chance that, if the entire population had been interviewed, the fraction of people supporting candidate X would be between 42 percent and 48 percent." Note that there is always some uncertainty in the results of a poll, which means that a poll cannot predict the winner of a very close election. Also note that there is no guarantee that the fraction of candidate X supporters in the population really is between 42 percent and 48 percent; there is a 5 percent chance that the true figure is outside that range. For an example of how to calculate the range of uncertainty, see **confidence interval.**

For another important topic in statistical inference, see **hypothesis testing.** For contrast, see **descriptive statistics.**

STATISTICS Statistics is the study of ways to analyze data. It consists of **descriptive statistics** and **statistical inference**. (Note that the word "statistics" is singular when it denotes the academic subject of statistics.)

STOCHASTIC A stochastic variable is the same as a random variable.

STOKES'S THEOREM Let **f** be a three-dimensional vector field, and let L be a continuous closed path. Stokes's theorem states that the line integral of **f** around L is equal to the surface integral of the curl of **f** around any surface S for which C is the boundary:

$$\int_{path\ C} \mathbf{f}(x,y,z) \cdot \mathbf{dL} = \int\int_{surface\ S} (\nabla \times \mathbf{F})\mathbf{dS}$$

This theorem is a generalization of Green's theorem, which applies to two dimensions. See **Green's theorem** for an example. For application, see **Maxwell's equations**. For background, see **line integral; surface integral; curl**.

SUBSCRIPT A subscript is a little number or letter set slightly below another number or letter. In the expression x_1, the "1" is a subscript.

SUBSET Set B is a subset of set A if every element contained in B is also contained in A. For example, the set of high school seniors is a subset of the set of all high school students. The set of squares is a subset of the set of rectangles, which in turn is a subset of the set of parallelograms. For illustration, see **Venn diagram**.

SUBSTITUTION PROPERTY The substitution property states that, if $a = b$, we can replace the expression a anywhere it appears by b if we wish. For example, in solving the simultaneous equation system $2x + 3y = 24$, $2y = 8$ we can solve the second equation to find $y = 4$, and then substitute 4 in place of y in the first equation:

$$2x + 3 \cdot 4 = 24$$

Therefore $x = 6$.

SUBTRACTION Subtraction is the opposite of addition. If $a + b = c$, then $c - b = a$. For example, $8 - 3 = 5$. Subtraction does not satisfy the commutative property:

$$a - b \neq b - a$$

nor the associative property:

$$(a - b) - c \neq a - (b - c)$$

SUFFICIENT In the statement "IF p, THEN q" $(p \rightarrow q)$, p is said to be a sufficient condition for q to be true. For example, being born in the United States is sufficient to become a United States citizen. (It is not necessary, though, because a person can become a naturalized citizen.) Showing that a number x is prime is sufficient to show that x is odd (if $x > 2$), but it is not necessary (for example, 9 is odd, but it is not prime).

SUM The sum is the result obtained when two numbers are added. In the equation $5 + 6 = 11$, 11 is the sum of 5 and 6.

SUMMATION NOTATION Summation notation provides a concise way of expressing long sums that follow a pattern. The Greek capital letter sigma $\sum$ is used to represent summation. Put where to start at the bottom:

$$\sum_{i=1}$$

and where to stop at the top:

$$\sum_{i=1}^{5}$$

and put what you want to add up along the sides:

$$\sum_{i=1}^{5} i$$

For example:

$$\sum_{i=1}^{5} i = 1 + 2 + 3 + 4 + 5 = 15$$

$$\sum_{j=1}^{10} j^2 = 1 + 4 + 9 + \ldots + 64 + 81 + 100$$

$$= 385$$

SUPPLEMENTARY Two angles are supplementary if the sum of their measures is 180°. For example, two angles measuring 135° and 45° form a pair of supplementary angles.

SURFACE A surface is a two-dimensional set of points. For example, a plane is an example of a surface; any point can be identified by two coordinates x and y. We live on the surface of the sphere formed by the Earth; any point can be identified by the two coordinates latitude and longitude.

SURFACE AREA The surface area of a solid is a measure of how much area the solid would have if you could somehow break it apart and flatten it out. For example, a cube with edge a units long has six faces, each with area a^2. The surface area of the cube is the sum of the areas of these six faces, or $6a^2$. The surface area of any polyhedron can be found by adding together the areas of all the faces. The surface areas of curved solids are harder to find, but they can often be found with calculus. (See **surface area, figure of revolution.**) Surface areas are important if you need to paint something. The amount of paint you need to completely paint an object depends on its surface area.

SURFACE AREA, FIGURE OF REVOLU-TION Suppose the curve $y = f(x)$ is rotated about the x-axis between the lines $x = a$ and $x = b$. (See figure 125.)

The surface area of this figure can be found with integration. Let dA represent the surface area of a small frustum cut from this figure. (See figure 126.)

The surface area of the frustum is $dA = 2\pi y ds$ where y is the average radius of the frustum, and ds is the slant height. ds is given by the formula $ds = \sqrt{1 + (\frac{dy}{dx})^2} dx$. (See **arc length.**) Then the total surface area is given by this integral:

$$\int_a^b 2\pi y \sqrt{1 + \left(\frac{dy}{dx}\right)^2}\, dx$$

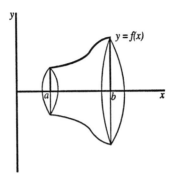

Figure 125   Surface formed by rotating $y = f(x)$ about x axis

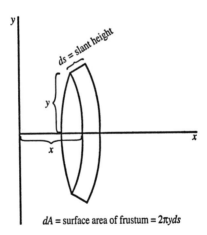

dA = surface area of frustum = $2\pi y ds$

Figure 126

For example, a sphere can be formed by rotating the curve $y = \sqrt{r^2 - x^2}$ about the x-axis from $x = -r$ to $x = r$. Then

$$\frac{dy}{dx} = \frac{-x}{\sqrt{r^2 - x^2}} = \frac{-x}{y}$$

The integral for the surface area is:

$$A = \int_{-r}^{r} 2\pi y \sqrt{1 + \frac{x^2}{y^2}}\, dx$$

$$A = \int_{-r}^{r} 2\pi \sqrt{x^2 + y^2}\, dx$$

$$A = 2\pi r \int_{-r}^{r} dx$$

$$A = 2\pi r x \big|_{-r}^{r}$$

$$A = 4\pi r^2$$

SURFACE INTEGRAL Let **E** be a three-dimensional vector field, and let **S** be a surface. Consider a small square on this surface. Create a vector **dS** whose magnitude is equal to the area of the small square, and whose direction is oriented to point outward along the surface. Calculate the dot product **E** · **dS**, and then integrate this dot product over the entire surface. The result is the surface integral of the field E along this surface:

$$(\text{surface integral}) = \iint_{surface} \mathbf{E} \cdot \mathbf{dS}$$

In order to evaluate the integral, the surface needs to expressed in terms of two parameters. The result will be a double integral, since the surface is two-dimensional. The example below is a simple case because the surface is a sphere.

Let $\mathbf{E}$ be vector field with magnitude given by:

$$\|\mathbf{E}\| = \frac{q}{4\pi\epsilon_0 r^2}$$

whose direction always point away from the origin. (This is the electric field created by a point electric charge with charge q located at the origin.)

Consider a surface integral along a sphere of radius r_0 centered at the charge. In this case the field vector $\mathbf{E}$ points in the same direction as the vector $\mathbf{dS}$, so the dot product between them will simply be the product of their magnitudes.

$$\mathbf{E} \cdot \mathbf{dS} = \|\mathbf{E}\| \times \|\mathbf{dS}\| = \frac{q}{4\pi\epsilon_0 r^2} dS$$

Since r is constant for a sphere, it can be pulled outside the integral, along with the other constants. The surface integral becomes:

$$\frac{q}{4\pi\epsilon_0 r^2} \iint_{sphere} dS$$

The double integral over the surface of the sphere just gives the surface area of the sphere, so the result is:

$$\frac{q}{4\pi\epsilon_0 r^2} 4\pi r^2 = \frac{q}{\epsilon_0}$$

For application, see **Maxwell's equations.**

SYLLOGISM In logic, a syllogism is a particular type of argument with three sentences: the major premise, which often asserts a general relationship between classes of objects; the minor premise, which asserts something about a specific case; and the conclusion, which follows from the two premises. Here is an example of a syllogism:

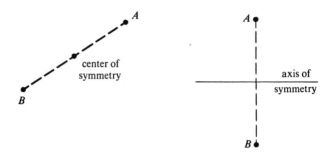

Figure 127

Major premise: All books about logic are interesting.

Minor premise: The *Dictionary of Mathematics Terms* is a book about logic.

Conclusion: Therefore, the *Dictionary of Mathematics Terms* is interesting.

SYMMETRIC (1) Two points A and B are symmetric with respect to a third point (called the *center of symmetry*) if the third point is the midpoint of the segment connecting the first two points. (See figure 127.)

(2) Two points A and B are symmetric with respect to a line (called the *axis of symmetry*) (see figure 127) if the line is the perpendicular bisector of the segment AB.

SYMMETRIC PROPERTY OF EQUALITY

The symmetric property of equality states that, if $a = b$, then $b = a$. That means that you can reverse the two sides of an equation whenever you want to.

SYNTHETIC DIVISION Synthetic division is a
short way of dividing a polynomial by a binomial
of the form $x - b$. For example, to find

$$\frac{3x^3 + 2x^2 - 165x + 28}{x - 7}$$

by algebraic division, we would have to write

$$
\begin{array}{r r r r}
 & 3x^2 & +23x & -4 \\
\hline
x - 7 \ )3x^3 & +2x^2 & -165x & +28 \\
3x^3 & -21x^2 & & \\
\hline
 & 23x^2 & -165x & \\
 & 23x^2 & -161x & \\
\hline
 & & -4x & +28 \\
 & & -4x & +28 \\
\hline
 & & & 0 \\
\end{array}
$$

To make synthetic division shorter, we leave out
all the x's and just write the coefficients. Also, we
reverse the sign of the divisor (so $(x - 7)$ becomes
$(x + 7)$ in this case) so as to make every intermediate subtraction become an addition. Finally, we
condense everything onto three lines. Here is a step-
by-step account: First, write the coefficients on a
line:

$$3 \quad 2 \quad -165 \quad 28 \quad)7$$

Second, bring down the first coefficient (3) into
the answer line:

$$3 \quad 2 \quad -165 \quad 28 \quad)7$$

$$\overline{}$$

$$3$$

Third, multiply the 3 in the answer by the 7 in the divisor, and write the result (21) on the second line as shown:

$$
\begin{array}{ccccc}
3 & 2 & -165 & 28 &)7 \\
  & 21 & & & \\
\hline
3 & & & &
\end{array}
$$

and then add:

$$
\begin{array}{ccccc}
3 & 2 & -165 & 28 &)7 \\
  & 21 & & & \\
\hline
3 & 23 & & &
\end{array}
$$

Now repeat the multiplication and addition procedure for the next two places:

$$
\begin{array}{ccccc}
3 & 2 & -165 & 28 &)7 \\
  & 21 & 161 & & \\
\hline
3 & 23 & -4 & &
\end{array}
$$

$$
\begin{array}{ccccc}
3 & 2 & -165 & 28 &)7 \\
  & 21 & 161 & -28 & \\
\hline
3 & 23 & -4 & 0 &
\end{array}
$$

The numbers in the answer line are, from left to right, the coefficients of x^2, x^1, and x^0. The farthest right entry in the answer line is the remainder (in this case 0). Therefore, the answer is $3x^2 + 23x - 4$.

The general procedure for synthetic division when the dividend is a third-degree polynomial:

$$\frac{a_3x^3 + a_2x^2 + a_1x + a_0}{x - b} = c_2x^2 + c_1x + c_0 + \frac{R}{x - b}$$

where the answer is found from:

a_3	a_2	a_1	a_0	$) b$
	$c_2 b$	$c_1 b$	$c_0 b$	
c_2	c_1	c_0	R	

The c's and R are defined as follows:

$$c_2 = a_3$$
$$c_1 = c_2 b + a_2$$
$$c_0 = c_1 b + a_1$$
$$R = c_0 b + a_0$$

SYSTEM OF EQUATIONS See **simultaneous equations.**

SYSTEM OF INEQUALITIES A system of inequalities is a group of inequalities that are all to be true simultaneously. For example, this system of three inequalities

$$x > 2$$
$$y > 3$$
$$x + y < 10$$

defines a set of values for x and y that will make all of the inequalities true. The graph of these points is shown in figure 128.

(See also **linear programming.**)

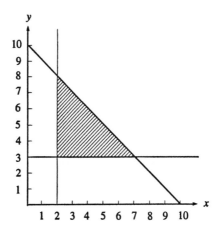

Figure 128   System of inequalities

T

t-DISTRIBUTION If Z is a random variable with a standard normal distribution, and Y has a chi-square (χ^2) distribution with n degrees of freedom (that is independent from Z) then the random variable

$$T = \frac{Z}{(Y/n)^{1/2}}$$

has a t-distribution with n degrees of freedom. $E(X)$ = 0 (if $n > 1$), and $\text{Var}(X) = n/(n-2)$ (if $n > 2$).

The density function for the t-distribution is bell-shaped, somewhat similar to the standard normal distribution. In fact, as n approaches infinity, the density function approaches the standard normal density function.

The t-distribution plays an important part in statistical estimation theory. (See **confidence interval**.) Tables 6 and 7 list some values for the t-distribution.

TANGENT (1) A tangent line is a line that intersects a circle at one point. Line AB in figure 129 is a tangent line. For example, the tires of a car are always tangent to the road. A tangent line to a curve is a line that just touches the curve, although it may intersect the curve at more than one point. For example, line CD in figure 129 is tangent to the curve at point E. The slope of a curve at any point is defined to be equal to the slope of the tangent line to the curve at that point. (See **calculus**.)

(2) If θ is an angle in a right triangle, then the tangent function in trigonometry is defined to be (opposite side)/(adjacent side). For an example of an

Figure 129 Tangent lines

application, suppose that you need to measure the height of a tall tree. It would be difficult to climb the tree with a tape measure, but you can walk 50 feet away from the tree and measure the angle of elevation of the top of the tree. (See figure 130.) If the angle is 55°, then you know that

$$\tan 55° = \frac{\text{(height of tree)}}{50}$$

Tan 55° = 1.43. This means that the height of the tree is $1.43 \times 50 = 71.5$ feet. This type of method is often used by surveyors when they need to measure the distance to faraway objects, and a similar type of method is used by astronomers to measure the distance to stars.

Tan θ is related to the other trigonometric functions by the equation:

$$\tan \theta = \frac{\sin \theta}{\cos \theta}$$

Here is a table of special values of the tangent function:

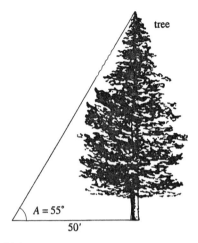

Figure 130 Finding height of tree with tangent function

θ (degrees)	θ (radians)	$\tan \theta$
0	0	0
30	$\pi/6$	$1/\sqrt{3}$
45	$\pi/4$	1
60	$\pi/3$	$\sqrt{3}$
90	$\pi/2$	infinity
180	π	0
270	$3\pi/2$	-infinity
360	2π	0

For most values of θ, $\tan \theta$ will be an irrational number. (See **trigonometry**.)

TAUTOLOGY A tautology is a sentence that is necessarily true because of its logical structure, regardless of the facts. For example, the sentence "The Earth is flat or else it is not flat" is a tautology. A tautology does not give you any information about the world, but studying the logical structure of tautologies is interesting. For example, let r represent the sentence

(p AND q) OR [(NOT p) OR (NOT q)]

The following truth table shows that the sentence r is a tautology:

p	q	p AND q	NOT p	NOT q	(NOT p) OR (NOT q)	r
T	T	T	F	F	F	T
T	F	F	F	T	T	T
F	T	F	T	F	T	T
F	F	F	T	T	T	T

All of the values in the last column are true. Therefore, r will necessarily be true, whether or not p or q is true. In words, sentence r says: "Either p and q are both true, or else at least one of them is not true."

The negation of a tautology is necessarily false; it is called a *contradiction*.

TAYLOR Brook Taylor (1685 to 1731) was a British mathematician who contributed to advances in calculus. (See **Taylor series.**)

TAYLOR SERIES The Taylor series expansion of a function $f(x)$ states that

$$f(x+h) = f(x) + hf'(x) + \frac{h^2 f''(x)}{2!} + \frac{h^3 f'''(x)}{3!} + \frac{h^4 f''''(x)}{4!} + \cdots$$

In this expression $f'(x)$ means the first derivative of f, $f''(x)$ means the second derivative, and so on.

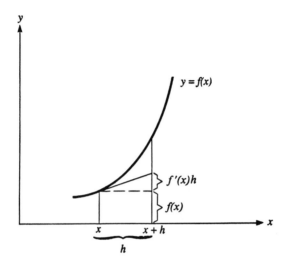

Figure 131

Taylor series are helpful when we know $f(x)$, but not $f(x + h)$. If the series goes on infinitely, we can often approximate the value of $f(x + h)$ by taking the first few terms of the series. By adding more and more terms we can make the approximation as close to the true value as we wish.

The first two terms of the series can be reached by approximating the curve by its tangent line. (See figure 131.)

For an example of where the additional terms come from, consider the third-degree polynomial function

$$f(x) = a_0 + a_1 x + a_2 x^2 + a_3 x^3$$

Then:

$$f(x + h) = a_0 + (a_1x + a_1h) + (a_2x^2 + 2a_2xh + a_2h^2)$$

$$+(a_3x^3 + 3a_3hx^2 + 3a_3xh^2 + a_3h^3)$$

$$= f(x) + (a_1h + 2a_2xh + 3a_3x^2h) + (a_2h^2 + 3a_3xh^2) + a_3h^3$$

$$= f(x) + h[a_1 + 2a_2x + 3a_3x^2] + \frac{h^2}{2}[2a_2 + 6a_3x] + \frac{h^3}{6}[6a_3]$$

By taking the values of the derivatives of f, we can see that

$$\begin{aligned} f'(x) &= a_1 + 2a_2x + 3a_3x^2 \\ f''(x) &= 2a_2 + 6a_3x \\ f'''(x) &= 6a_3 \end{aligned}$$

Therefore, in this case:

$$f(x + h) = f(x) + hf'(x) + \frac{h^2 f''(x)}{2!} + \frac{h^3 f'''(x)}{3!}$$

and no higher terms are needed in the series.

Taylor series make it possible to find expressions to calculate some functions, such as $\sin\theta$. Since $\sin\theta = \sin(0 + \theta)$, we can form the Taylor expansion:

$$\sin\theta = \sin 0 + \theta\cos 0 - \frac{\theta^2 \sin 0}{2} - \frac{\theta^3 \cos 0}{3!} + \frac{\theta^4 \sin 0}{4!} + \cdots$$

(using the fact that $d\sin\theta/d\theta = \cos\theta$, and $d\cos\theta/d\theta = -\sin\theta$). ($\theta$ is in radians.)

Since $\sin 0 = 0$ and $\cos 0 = 1$, we have

$$\sin\theta = \theta - \frac{\theta^3}{3!} + \frac{\theta^5}{5!} - \frac{\theta^7}{7!} + \frac{\theta^9}{9!} - \cdots$$

Other examples of Taylor series are as follows:

$$\cos\theta = 1 - \frac{\theta^2}{2!} + \frac{\theta^4}{4!} - \frac{\theta^6}{6!} + \cdots$$

$$e^x = 1 + x + \frac{x^2}{2!} + \frac{x^3}{3!} + \frac{x^4}{4!} + \cdots$$

TERM A term is a part of a sum. For example, in the polynomial $ax^2 + bx + c$, the first term is ax^2, the second term is bx, and the third term is c. The different terms in an expression are separated by addition (or subtraction) signs.

TERMINAL SIDE When discussing general angles in trigonometry, it is convenient to place the vertex of the angle at the origin and to orient the angle in such a way that one side points along the positive x-axis. Then the other side of the angle is said to be the terminal side.

TERMINATING DECIMAL A terminating decimal is a fraction whose decimal representation contains a finite number of digits. For example, $\frac{1}{4} = 0.25$, and $\frac{5}{32} = 0.15625$. For contrast, see **repeating decimal.**

TEST STATISTIC A test statistic is a quantity calculated from observed sample values that is used to test a null hypothesis. The test statistic is constructed so that it will come from a known distribution if the null hypothesis is true. Therefore, the null hypothesis is rejected if it seems implausible that the observed value of the test statistic could have come from that distribution. (See **hypothesis testing.**)

TETRAHEDRON A tetrahedron is a polyhedron with four faces. Each face is a triangle. In other

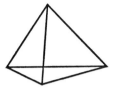

Figure 132   Tetrahedron

words, a tetrahedron is a pyramid with a triangular base. A regular tetrahedron has all four faces congruent. (See figure 132.)

THEN The word "THEN" is used as a connective word in logic sentences of the form "$p \rightarrow q$" ("IF p, THEN q.") Here is an example: "If a triangle has three equal sides, then it has three equal angles."

THEOREM A theorem is a statement that has been proved, such as the Pythagorean theorem.

TOROID A toroid can be formed by rotating a closed curve for a full turn about a line that is in the same plane as the curve, but does not cross it. The set of all points that the curve crosses in the course of the rotation forms a toroid.

TORUS A torus is a solid figure formed by rotating a circle about a line in the same plane as the circle, but not on the circle. A doughnut is an example of a torus.

TRACE The trace of a square matrix is the sum of the diagonal elements of the matrix. For example, the trace of

$$\begin{pmatrix} 1 & 2 & 9 \\ 7 & 3 & 4 \\ 8 & 5 & 6 \end{pmatrix}$$

is equal to: $1 + 3 + 6 = 10$.

TRAJECTORY The trajectory is the path that a body makes as it moves through space.

TRANSCENDENTAL NUMBER A transcendental number is a number that cannot occur as the root of a polynomial equation with rational coefficients. The transcendental numbers are a subset of the irrational numbers. Most values for trigonometric functions are transcendental, as is the number e. The number π is transcendental, but this fact was not proved until 1882. The square roots of rational numbers are not transcendental, even though they are irrational. For example, $\sqrt{6}$ is a root of the equation $x^2 - 6 = 0$, so it is not transcendental.

TRANSITIVE PROPERTY The transitive property of equality states that, if $a = b$ and $b = c$, then $a = c$. All real and complex numbers obey this property.

   The transitive property of inequality states that, if $a > b$ and $b > c$, then $a > c$. Real numbers obey this property, but complex numbers do not.

TRANSLATION A translation occurs when we shift the axes of a Cartesian coordinate system. (See figure 133.) (We keep the orientation of the axes the same; otherwise there would be a rotation.) If the new coordinates are called x' and y' (x-prime and

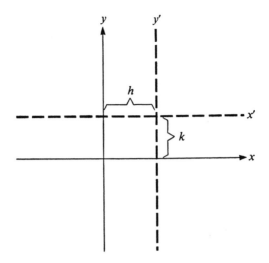

Figure 133 Translation of coordinate axes

y-prime), and the amount that the x-axis is shifted is h and the amount that the y-axis is shifted is k, then there is a simple relation between the new coordinates and the old coordinates:

$$x' = x - h$$
$$y' = y - k$$

TRANSPOSE The transpose of a matrix is formed by turning all the columns in the original matrix into rows in the transposed matrix. For example:

$$\begin{pmatrix} a & b \\ c & d \end{pmatrix}^{tr} = \begin{pmatrix} a & c \\ b & d \end{pmatrix}$$

$$\begin{pmatrix} 1 & 2 & 3 \\ 4 & 5 & 6 \end{pmatrix}^{tr} = \begin{pmatrix} 1 & 4 \\ 2 & 5 \\ 3 & 6 \end{pmatrix}$$

If a matrix **A** has m rows and n columns, then $\mathbf{A}^{tr}$ will have n rows and m columns.

TRANSVERSAL A transversal is a line that intersects two lines. For examples, see **corresponding angles** and **alternate interior angles**.

TRAPEZOID A trapezoid is a quadrilateral that has exactly two sides parallel. For illustration, see **quadrilateral**.

TREE DIAGRAM A tree diagram illustrates all of the possible results for a process with several stages. Figure 134 illustrates a tree diagram that shows all of the possible results for tossing three coins.

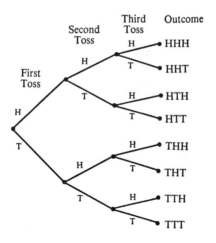

Figure 134   Tree diagram

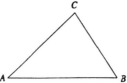

Figure 135   Triangle

TRIANGLE A triangle is a three-sided polygon. (See figure 135.) The three points where the sides intersect are called *vertices*. Triangles are sometimes identified by listing their vertices, as in triangle *ABC*.

One reason that triangles are important is that they are rigid. If you imagine the three sides of a triangle as joined by hinges, you could not bend the triangle out of shape. However, you could easily bend a quadrilateral or any other polygon out of shape if its vertices were formed with hinges. Triangle-shaped supports are often used in bridge construction.

If you add together the three angles in any triangle, the result will be 180°. To prove this, draw line *DE* parallel to line *AC*, as in figure 136. Then angle 1 = angle 2, and angle 4 = angle 5, since they are alternate interior angles between parallel lines. We can also see that angle 2 + angle 3 + angle 4 = 180°, since *DBE* is a straight line. Then, by substitution, angle 1 + angle 3 + angle 5 = 180°.

The area of a triangle is equal to $\frac{1}{2}$(base)(altitude), where (base) is the length of one of the sides, and (altitude) is the perpendicular distance from the base to the opposite vertex. (See figure 137.)

If one of the three angles in a triangle is an obtuse angle, the triangle is called an *obtuse triangle*. If each of the three angles is less than 90°, it is called an *acute triangle*. If one angle equals 90°, it is called

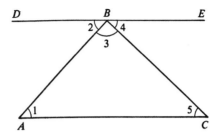

Figure 136

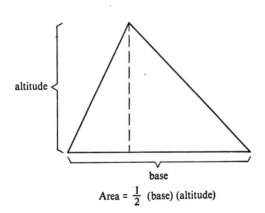

Area = $\frac{1}{2}$ (base) (altitude)

Figure 137

a *right triangle*.

If the three sides of a triangle are equal, it is called an *equilateral triangle*. If two sides are equal, it is called an *isosceles triangle*. Otherwise, it is a *scalene triangle*.

Two triangles are *congruent* if they have the same shape and size. There are several ways to show that triangles are congruent:

(1) Side-side-side: Two triangles are congruent if all three of their corresponding sides are equal.

(2) Side-angle-side: Two triangles are congruent if two corresponding sides and the angle between them are equal.

(3) Angle-side-angle: Two triangles are congruent if two corresponding angles and the side between them are equal.

(4) Angle-angle-side: Two triangles are congruent if two corresponding angles and any corresponding side are equal.

(5) Leg-hypotenuse: Two right triangles are congruent if the hypotenuse and two corresponding legs are equal.

If all three of the angles of the two triangles are equal, then the triangles have exactly the same shape. However, they may not have the same size. For example, the White House, the Capitol, and the Washington Monument form a triangle, and the marks representing these three buildings on a map also form a triangle. The two triangles have the same shape, so they are said to be *similar*, but the triangle formed by the real buildings is clearly much bigger than the triangle formed by the marks on the map. The corresponding sides of similar triangles are in proportion (meaning that, if one side of the big

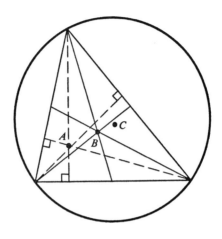

— — —: altitudes
————: medians
Point A: orthocenter
Point B: centroid
Point C: circumcenter

Figure 138

triangle is 10 times as large as the corresponding side
on the little triangle, then the other two sides on the
big triangle will also be 10 times as large as their
corresponding sides on the little triangle).

A line segment that joins the vertex of a triangle
to the midpoint of the opposite side is called a *median*. The point where the three medians intersect is
called the *centroid*; it is the point where the triangle
would balance if supported at a single point. The
point where the three altitudes of the triangle join is
called the *orthocenter*. The point where the perpendicular bisectors of the three sides cross is called the

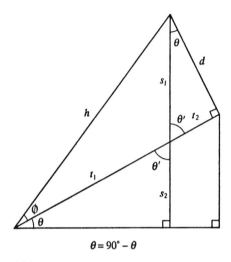

$\theta = 90° - \theta$

Figure 139

circumcenter; it is the center of the circle that can be circumscribed about that triangle. (See figure 138.) For illustration of the circle that can be inscribed in a triangle, see **incircle.**

TRIGONOMETRIC FUNCTIONS OF A SUM

Suppose that we need to find $\sin(\theta + \phi)$, where θ and ϕ are two angles as shown in figure 139. We can see that

$$\sin(\theta + \phi) = \frac{s_1 + s_2}{h}$$

We can find s_2 from the equation

$$s_2 = t_1 \sin \theta$$

We can find t_1:

$$\frac{t_1 + t_2}{h} = \cos \phi; \quad \text{then} \quad t_1 = h \cos \phi - t_2$$

Now we find t_2 from

$$t_2 = s_1 \sin \theta$$

and put this value for t_2 back in the equation for t_1:

$$t_1 = h \cos \phi - s_1 \sin \theta$$

Putting this expression back in the equation for s_2 gives

$$s_2 = h \cos \phi \sin \theta - s_1 \sin^2 \theta$$

Putting this expression back in the equation for $\sin(\theta + \phi)$ we obtain

$$
\begin{aligned}
\sin(\theta + \phi) &= \frac{1}{h}[s_1 + h \cos \phi \sin \theta - s_1 \sin^2 \theta] \\
&= \frac{1}{h}[s_1(1 - \sin^2 \theta) + h \cos \phi \sin \theta] \\
&= \frac{1}{h}[s_1 \cos^2 \theta + h \cos \phi \sin \theta] \\
&= \frac{s_1 \cos^2 \theta}{h} + \cos \phi \sin \theta
\end{aligned}
$$

From the definitions of the trigonometric functions, we know that:

$$h = \frac{d}{\sin \phi}, \quad s_1 = \frac{d}{\cos \theta}, \quad \frac{s_1}{h} = \frac{\sin \phi}{\cos \theta}$$

The final formula becomes:

$$\sin(\theta + \phi) = \sin \phi \cos \theta + \sin \theta \cos \phi$$

From this formula we can derive a similar formula for cosine:

$$
\begin{aligned}
\cos(\theta + \phi) &= \sin(90° - \theta - \phi) \\
&= \sin(90° - \theta)\cos(-\phi) + \cos(90° - \theta)\sin(-\phi) \\
&= \cos \theta \cos \phi - \sin \theta \sin \phi
\end{aligned}
$$

and a formula for tangent:

$$
\begin{aligned}
\tan(\theta + \phi) &= \frac{\sin(\theta + \phi)}{\cos(\theta + \phi)} \\
&= \frac{\sin\theta\cos\phi + \sin\phi\cos\theta}{\cos\theta\cos\phi - \sin\theta\sin\phi} \\
&= \frac{\dfrac{\sin\theta\cos\phi}{\cos\theta\cos\phi} + \dfrac{\sin\phi\cos\theta}{\cos\theta\cos\phi}}{\dfrac{\cos\theta\cos\phi}{\cos\theta\cos\phi} - \dfrac{\sin\theta\sin\phi}{\cos\theta\cos\phi}} \\
\tan(\theta + \phi) &= \frac{\tan\theta + \tan\phi}{1 - \tan\theta\tan\phi}
\end{aligned}
$$

We can find double-angle formulas by setting $\theta = \phi$:

$$
\begin{aligned}
\sin 2\theta &= 2\sin\theta\cos\theta \\
\cos 2\theta &= \cos^2\theta - \sin^2\theta = 1 - 2\sin^2\theta = 2\cos^2\theta - 1 \\
\tan 2\theta &= \frac{2\tan\theta}{1 - \tan^2\theta}
\end{aligned}
$$

TRIGONOMETRY Trigonometry is the study of triangles. In particular, six functions are called the trigonometric functions: sine, cosine, tangent, cotangent, secant, and cosecant. Although these functions were originally developed to help solve problems involving triangles, it turns out that they have many other applications.

Trigonometric functions can be illustrated by considering a circle of radius r centered at the origin. Draw an angle θ with vertex at the origin and initial side along the x axis. Then, let (x, y) be the coordinates of the point where the terminal side of the angle crosses the circle. The definitions of the trigonometric functions are:

$$
\sin\theta = \frac{y}{r} \qquad \csc\theta = \frac{r}{y}
$$

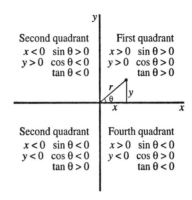

Figure 140

$$\cos \theta = \frac{x}{r} \qquad \sec \theta = \frac{r}{x}$$
$$\tan \theta = \frac{y}{x} \qquad \text{ctn}\theta = \frac{x}{y}$$

Whether the value of a trigonometric function is positive or negative depends upon the quadrant. Figure 140 shows the sign of the value for sin, cos, and tan for each of the four quadrants.

An angle is completely unchanged if we add 2π radians to it. This means that

$$\sin \theta = \sin(\theta + 2\pi) = \sin(\theta + 4\pi) = \sin(\theta + 6\pi) = \ldots$$

Therefore the trigonometric functions are periodic, or cyclic. For every 2π units, they will have the same value.

Figure 141 shows the graphs of the sine, cosine, and tangent functions. The sine function can be used to describe many types of periodic motion. The curve describes the motion of a weight attached to a spring or a swinging pendulum. It describes the voltage change with time in an alternating-current circuit with a rotating generator. The movement of the tides is approximately sine-shaped, as is the variation of the length of the day throughout the year. The sine function is also used to describe light waves, water waves, and sound waves. Table 2 lists some values for trigonometric functions.

(See also **trigonometric functions of a sum; inverse trigonometric functions.**)

TRINOMIAL A trinomial is the indicated sum of three monomials. For example, $10 + 13x^2 + 20a^3b^2$ is a trinomial.

TRIPLE INTEGRAL A triple integral means to integrate a function over an entire volume. For example, if $\rho(x, y, z)$ represents the density of matter at a point (x, y, z), then

$$\int_{z=0}^{z=c} \int_{y=0}^{y=b} \int_{x=0}^{x=a} \rho(x,y,z) \, dx \, dy \, dz$$

gives the total mass contained in the parallelepiped from $x = 0$ to $x = a$, $y = 0$ to $y = b$, and $z = 0$ to $z = c$.

TRISECT To trisect an object means to cut it in three equal parts. For example, one can trisect a line segment, or trisect an angle. (See **geometric construction.**)

TRUE "True" is one of the two truth values attached to sentences in logic. It corresponds to what we normally suppose: "true" means "accurate," "correct." (See **logic; Boolean algebra.**)

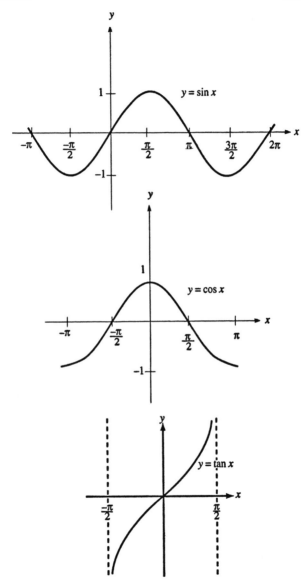

Figure 141

TRUNCATED CONE A truncated cone consists of the section of a cone between the base and another plane that intersects the cone between the base and the vertex. It looks like a cone whose top has been chopped off.

TRUNCATED PYRAMID A truncated pyramid consists of the section of a pyramid between the base and another plane that intersects the pyramid between the base and the vertex. It looks like a pyramid whose top has been chopped off.

TRUNCATION The truncation of a number is found by dropping the fractional part of that number. It is equal to the largest integer that is less than or equal to the original number. For example, the truncation of 17.89 is equal to 17.

TRUTH TABLE A truth table is a table showing whether a compound logic sentence will be true or false, based on whether the simple sentences contained in the compound sentence are true. Each row of the table corresponds to one set of possible truth values for the simple sentences. For example, if there are three simple sentences, then there will be $2^3 = 8$ rows in the truth table. Here is a truth table that demonstrates De Morgan's law:

NOT (p OR q)
is equivalent to
(NOT p) AND (NOT q).

p	q	p OR q	NOT (p OR q)	NOT p	NOT q	(NOT p) AND (NOT q)
T	T	T	F	F	F	F
T	F	T	F	F	T	F
F	T	T	F	T	F	F
F	F	F	T	T	T	T

The first two columns contain the simple sentences p and q. Since there are four possible combinations of truth values for p and q, the table contains four rows. Each of the five remaining columns tells us whether the indicated expression will be true or false, given the possible values for p and q. Note that the column for NOT $(p$ OR $q)$ and the column for (NOT p) AND (NOT q) have exactly the same values, so these two sentences are equivalent.

TRUTH VALUE In logic, a sentence is assigned one of two truth values. One of the truth values is labeled T, or 1; it corresponds to "true." The other truth value is labeled F, or 0; it corresponds to "false." The question "What does it mean for a sentence to be true?" is a very difficult philosophical question. In logic a sentence is said to have the truth value T or F, rather than to be "true" or "false"; this makes it possible to analyze the validity of arguments containing "true" or "false" sentences without having to answer the question as to what "truth" really means.

TWO-TAILED TEST In a two-tailed test the critical region consists of both tails of a distribution. The null hypothesis is rejected if the test-statistic value is either too large or too small. (See **hypothesis testing**.)

TYPE 1 ERROR A type 1 error occurs when the null hypothesis is rejected when it is actually true. (See **hypothesis testing**.)

TYPE 2 ERROR A type 2 error occurs when the null hypothesis is accepted when it is actually false. (See **hypothesis testing**.)

U

UNBIASED ESTIMATOR An unbiased estimator is an estimator whose expected value is equal to the true value of the parameter it is trying to estimate. (See **statistical inference**.)

UNDEFINED TERM An undefined term is a basic concept that is described, rather than given a rigorous definition. It would be impossible to rigorously define every term, because sooner or later the definitions would become circular. "Line" is an example of an undefined term from geometry.

UNION The union of two sets A and B (written as $A \cup B$) is the set of all elements that are either members of A or members of B, or both. For example, the union of the sets $A = \{0, 1, 2, 3, 4\}$ and $B = \{2,4,6,8,10,12\}$ is the set $A \cup B = \{0,1,2,3,4,6,8,10,12\}$ The union of the set of whole numbers and the set of negative integers is the set of all integers.

UNIT VECTOR A unit vector is a vector of length 1. It is common to use **i** to represent the unit vector along the x axis—that is, the vector whose components are $(1,0,0)$. Likewise, **j** is used to represent $(0,1,0)$, and **k** represents $(0,0,1)$. A three-dimensional vector whose components are (x, y, z) can be written as the vector sum of each component times the corresponding unit vector: $(x, y, z) = x\mathbf{i} + y\mathbf{j} + z\mathbf{k}$.

UNIVERSAL QUANTIFIER An upside-down letter A, $\forall$, is used to represent the expression "For all ...," and is called the universal quantifier. For example, if x is allowed to take on real-number values, then the sentence "For all real numbers, the square of the number is nonnegative" can be written as

(1) $\forall_x(x^2 \geq 0)$

For another example, let C_x, represent the sentence "x is a cow," and let M_x represent the sentence "x says moo." Then the expression

(2) $\forall_x(C_x \rightarrow M_x)$

represents the sentence "For all x, if x is a cow, then x says moo." In more informal terms, the sentence could be written as "All cows say moo."

Be careful when taking the negation of a sentence that uses the universal quantifier. The negation of sentence (2) is not the sentence "All cows do not say moo," which would be written as

(3) $\forall_x(C_x \rightarrow \text{NOT}M_x)$

Instead, the negation of sentence (2) is the sentence "Not all cows say moo," which can be written as

(4) $NOT\forall_x(C_x \rightarrow M_x)$

Sentence (4) could also be written as

(5) $\exists_x(C_x \text{AND NOT}M_x)$

(See **existential quantifier**.)

UNIVERSAL SET The universal set is the set of all objects in which you are interested during a particular discussion. For example, in talking about numbers the relevant universal set might be the set of all complex numbers.

V

VARIABLE A variable is a symbol that is used to represent a value from a particular set. For example, in algebra it is common to use letters to represent values from the set of real numbers. (See **algebra**.)

VARIANCE The variance of a random variable X is defined to be

$$Var(X) = E[(X - E(X)) \times (X - E(X))]$$
$$= E[(X - E(X))^2]$$

where E stands for "expectation."

The variance can also be found from the formula:

$$Var(X) = E(X^2) - [E(X)]^2$$

The variance is often written as σ^2. (The Greek lower-case letter sigma (σ), is used to represent the square root of the variance, known as the *standard deviation*.)

The variance is a measure of how widespread the observations of X are likely to be. If you know for sure what the value of X will be, then $Var(X) = 0$.

For example, if X is the number of heads that appear when a coin is tossed five times, then the probabilities are given in this table:

i	$\mathbf{Pr}(X = i)$	$i \times \mathbf{Pr}(X = i)$	$i^2 \times \mathbf{Pr}(X = i)$
0	1/32	0	0
1	5/32	5/32	5/32
2	10/32	20/32	40/32
3	10/32	30/32	90/32
4	5/32	20/32	80/32
5	1/32	5/32	25/32
sum:	1	2.5	7.5

The sum of column 3 $[i \times \Pr(X = i)]$ gives $E(X) = 2.5$; the sum of column 4 $[i^2 \times \Pr(X = i)]$ gives $E(X^2) = 7.5$. From this information we can find

$$Var(X) = E(X^2) - [E(X)]^2 = 7.5 - 2.5^2 = 1.25$$

Some properties of the variance are as follows. If a and b are constants:

$$Var(aX + b) = a^2 Var(X)$$

If X and Y are independent random variables:

$$Var(X + Y) = Var(X) + Var(Y)$$

In general:

$$Var(X + Y) = Var(X) + Var(Y) + 2Cov(X, Y)$$

where $Cov(X, Y)$ is the covariance.

The variance of a list of numbers $x_1, x_2, \ldots x_n$ is given by either of these formulas:

$$
\begin{aligned}
Var(x) &= \frac{(x_1 - \bar{x})^2 + (x_2 - \bar{x})^2 + \cdots + (x_n - \bar{x})^2}{n} \\
&= \overline{x^2} - (\bar{x})^2
\end{aligned}
$$

where a bar over a quantity signifies average.

VECTOR A vector is a quantity that has both magnitude and direction. The quantity "60 miles per hour" is a regular number, or scalar. The quantity "60 miles per hour to the northwest" is a vector, because it has both size and direction. Vectors can be represented by drawing pictures of them. A vector is drawn as an arrow pointing in the direction of the vector, with length proportional to the size of the vector. (See figure 142.)

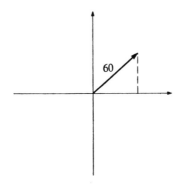

Figure 142 Vector

Vectors can also be represented by an ordered list of numbers, such as (3,4) or (1, 0, 3). Each number in this list is called a *component* of the vector. A vector in a plane (two dimensions) can be represented as an ordered pair. A vector in space (three dimensions) can be represented as an ordered triple.

Vectors are symbolized in print by boldface type, as in "vector **a**." A vector can also be symbolized by placing an arrow over it: $\vec{a}$.

The length, or magnitude, of a vector **a** is written as $\|\mathbf{a}\|$

Addition of vectors is defined as follows: Move the tail of the second vector so that it touches the head of the first vector, and then the sum vector (called the *resultant*) stretches from the tail of the first vector to the head of the second vector. (See figure 143.) For vectors expressed by components, addition is easy: just add the components:

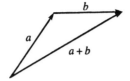

Figure 143 Adding vectors

$$(3,2) + (4,1) = (7,3)$$
$$(a,b) + (c,d) = (a+c, b+d)$$

To multiply a scalar by a vector, multiply each component by that scalar:

$$10(3,2) = (30,20)$$
$$n(a,b) = (na, nb)$$

To find two different ways of multiplying vectors, see **dot product** and **cross product**.

VECTOR FIELD A two-dimensional vector field **f** transforms a vector (x,y) into another vector $\mathbf{f}(x,y) = [f_x(x,y),\ f_y(x,y)]$. Here $f_x(x,y)$ and $f_y(x,y)$ are the two components of the vector field; each is a scalar function of two variables. An example of a vector field is:

$$\mathbf{f}(x,y) = \left[\frac{-y}{\sqrt{x^2 + y^2}},\ \frac{x}{\sqrt{x^2 + y^2}} \right]$$

If we evaluate this vector field at (3,4) we find:

$$\mathbf{f}(3,4) = \left[\frac{-4}{5},\ \frac{3}{5} \right]$$

In this particular case, the output of the vector field is perpendicular to the input vector.

The same concept can be generalized to higher-dimensional vector fields. For examples of calculus operations on vector fields, see **divergence; curl; line integral; surface integral; Stokes' theorem; Maxwell's equations.**

VECTOR PRODUCT This is a synonym for **cross product.**

VELOCITY The velocity vector represents the rate of change of position of an object. To specify a velocity, it is necessary to specify both a speed and a direction (for example, 50 miles per hour to the northwest).

If the motion is in one dimension, then the velocity is the derivative of the function that gives the position of the object as a function of time. The derivative of the velocity is called the **acceleration.**

If the vector $[x(t), y(t), z(t)]$ gives the position of the object in three dimensional space, where each component of the vector is given as a function of time, then the velocity vector is the vector of derivatives of each component:

$$\text{velocity} = [\frac{dx}{dt}, \frac{dy}{dt}, \frac{dz}{dt}]$$

VENN DIAGRAM A Venn diagram (see figure 144) is a picture that illustrates the relationships between sets. The universal set you are considering is represented by a rectangle, and sets are represented by circles or ellipses. The possible relationships between two sets A and B are as follows:

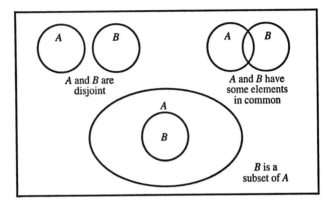

Figure 144

Set B is a subset of set A, or set A is a subset of set B,

Set A and set B are disjoint (they have no elements in common).

Set A and set B have some elements in common.

Figure 145 is a Venn diagram for the universal set of complex numbers.

VERTEX The vertex of an angle is the point where the two sides of the angle intersect.

VERTICAL ANGLES Two pairs of vertical angles are formed when two lines intersect. In figure 146, angle 1 and angle 2 are a pair of vertical angles. Angle 3 and angle 4 are another pair of vertical angles. The two angles in a pair of vertical angles are always equal to each other.

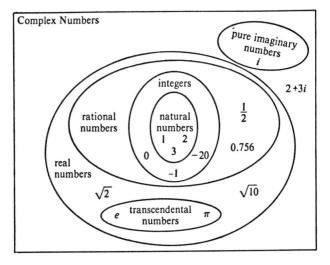

Figure 145   Venn diagram

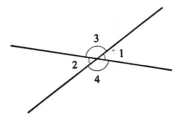

Figure 146   Vertical angles

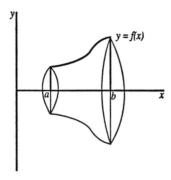

Figure 147 Surface formed by rotating $y = f(x)$ about x axis

VOLUME The volume of a solid is a measure of how much space it occupies. The volume of a cube with edge a units long is a^3. Volumes of other solids are measured in cubic units. The volume of a prism or cylinder is (base area) × (altitude), and the volume of a pyramid or cone is $(1/3)$ × (base area) × (altitude). (See also **volume, figure of revolution.**)

VOLUME, FIGURE OF REVOLUTION Suppose the curve $y = f(x)$ is rotated about the x-axis between the lines $x = a$ and $x = b$. (See figure 147.)

The volume of this figure can be found with integration. Let dV represent the volume of a small cylinder cut from this figure. (See figure 148.)

$$dV = \pi y^2 dx$$

where y is the radius of the cylinder, and dx is the height of the cylinder.

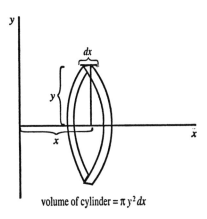

volume of cylinder = $\pi y^2 dx$

Figure 148

The volume of the entire figure is given by this integral:

$$V = \int_a^b \pi y^2 dx$$

For example, a sphere can be formed by rotating the circle $y = \sqrt{r^2 - x^2}$ about the x-axis from $x = -r$ to $x = r$. The volume is given by the integral:

$$
\begin{aligned}
V &= \int_{-r}^r \pi(r^2 - x^2)dx \\
&= \pi(r^2 x - \frac{1}{3}x^3)\big|_{-r}^r \\
&= \pi[r^3 - \frac{1}{3}r^3 - (-r^3 - \frac{1}{3}(-r)^3)] \\
&= \pi[2r^3 - \frac{2}{3}r^3] \\
V &= \frac{4}{3}\pi r^3
\end{aligned}
$$

W

WELL-FORMED FORMULA A well-formed formula (or *wff*) is a sequence of symbols that is an acceptable formula in logic. For example, the sequence *p* AND *q* is a *wff*, but the sequence AND *pq* is not a *wff*.

Certain rules govern the formation of *wff*'s in a particular type of logic. Here is an example of such a rule: If *p* and *q* are *wff*'s, then (*p* AND *q*) is also a *wff*.

WHOLE NUMBERS The set of whole numbers includes zero and all the natural numbers 0, 1, 2, 3, 4, 5, 6, ...

X

X-AXIS he *x*-axis is the horizontal axis in a Cartesian coordinate system.

X-INTERCEPT The *x*-intercept of a curve is the value of *x* at the point where the curve crosses the *x* axis.

Y

Y-AXIS The *y*-axis is the vertical axis in a Cartesian coordinate system.

Y-INTERCEPT The *y*-intercept of a curve is the value of *y* at the point where the curve crosses the *y*-axis.

Z

ZERO Intuitively, zero means nothing—for example, the score that each team has at the beginning of a game is zero. Formally, zero is the identity element for addition, which means that, if you add zero to any number, the number remains unchanged. In our number system the symbol "0" also serves as a place holder in the decimal representation of a number. Without zero we would have trouble telling the difference between 1000 and 10. Historically, the use of zero as a placeholder preceded the use of zero as a number in its own right.

APPENDIX

TABLE 1

Common Logarithm Table
The table gives log $(a + b)$.

a b:	.00	.01	.02	.03	.04	.05	.06	.07	.08	.09
1.0	.0000	.0043	.0086	.0128	.0170	.0212	.0253	.0294	.0334	.0374
1.1	.0414	.0453	.0492	.0531	.0569	.0607	.0645	.0682	.0719	.0755
1.2	.0792	.0828	.0864	.0899	.0934	.0969	.1004	.1038	.1072	.1106
1.3	.1139	.1173	.1206	.1239	.1271	.1303	.1335	.1367	.1399	.1430
1.4	.1461	.1492	.1523	.1553	.1584	.1614	.1644	.1673	.1703	.1732
1.5	.1761	.1790	.1818	.1847	.1875	.1903	.1931	.1959	.1987	.2014
1.6	.2041	.2068	.2095	.2122	.2148	.2175	.2201	.2227	.2253	.2279
1.7	.2304	.2330	.2355	.2380	.2405	.2430	.2455	.2480	.2504	.2529
1.8	.2553	.2577	.2601	.2625	.2648	.2672	.2695	.2718	.2742	.2765
1.9	.2788	.2810	.2833	.2856	.2878	.2900	.2923	.2945	.2967	.2989
2.0	.3010	.3032	.3054	.3075	.3096	.3118	.3139	.3160	.3181	.3201
2.1	.3222	.3243	.3263	.3284	.3304	.3324	.3345	.3365	.3385	.3404
2.2	.3424	.3444	.3464	.3483	.3502	.3522	.3541	.3560	.3579	.3598
2.3	.3617	.3636	.3655	.3674	.3692	.3711	.3729	.3747	.3766	.3784
2.4	.3802	.3820	.3838	.3856	.3874	.3892	.3909	.3927	.3945	.3962
2.5	.3979	.3997	.4014	.4031	.4048	.4065	.4082	.4099	.4116	.4133
2.6	.4150	.4166	.4183	.4200	.4216	.4232	.4249	.4265	.4281	.4298
2.7	.4314	.4330	.4346	.4362	.4378	.4393	.4409	.4425	.4440	.4456
2.8	.4472	.4487	.4502	.4518	.4533	.4548	.4564	.4579	.4594	.4609
2.9	.4624	.4639	.4654	.4669	.4683	.4698	.4713	.4728	.4742	.4757
3.0	.4771	.4786	.4800	.4814	.4829	.4843	.4857	.4871	.4886	.4900
3.1	.4914	.4928	.4942	.4955	.4969	.4983	.4997	.5011	.5024	.5038
3.2	.5052	.5065	.5079	.5092	.5105	.5119	.5132	.5145	.5159	.5172
3.3	.5185	.5198	.5211	.5224	.5237	.5250	.5263	.5276	.5289	.5302
3.4	.5315	.5328	.5340	.5353	.5366	.5378	.5391	.5403	.5416	.5428
3.5	.5441	.5453	.5465	.5478	.5490	.5502	.5515	.5527	.5539	.5551
3.6	.5563	.5575	.5587	.5599	.5611	.5623	.5635	.5647	.5658	.5670
3.7	.5682	.5694	.5705	.5717	.5729	.5740	.5752	.5763	.5775	.5786
3.8	.5798	.5809	.5821	.5832	.5843	.5855	.5866	.5877	.5888	.5899
3.9	.5911	.5922	.5933	.5944	.5955	.5966	.5977	.5988	.5999	.6010
4.0	.6021	.6031	.6042	.6053	.6064	.6075	.6085	.6096	.6107	.6117
4.1	.6128	.6138	.6149	.6160	.6170	.6180	.6191	.6201	.6212	.6222
4.2	.6232	.6243	.6253	.6263	.6274	.6284	.6294	.6304	.6314	.6325
4.3	.6335	.6345	.6355	.6365	.6375	.6385	.6395	.6405	.6415	.6425
4.4	.6435	.6444	.6454	.6464	.6474	.6484	.6493	.6503	.6513	.6522
4.5	.6532	.6542	.6551	.6561	.6571	.6580	.6590	.6599	.6609	.6618
4.6	.6628	.6637	.6646	.6656	.6665	.6675	.6684	.6693	.6702	.6712
4.7	.6721	.6730	.6739	.6749	.6758	.6767	.6776	.6785	.6794	.6803
4.8	.6812	.6821	.6830	.6839	.6848	.6857	.6866	.6875	.6884	.6893
4.9	.6902	.6911	.6920	.6928	.6937	.6946	.6955	.6964	.6972	.6981

a	b: .00	.01	.02	.03	.04	.05	.06	.07	.08	.09
5.0	.6990	.6998	.7007	.7016	.7024	.7033	.7042	.7050	.7059	.7067
5.1	.7076	.7084	.7093	.7101	.7110	.7118	.7126	.7135	.7143	.7152
5.2	.7160	.7168	.7177	.7185	.7193	.7202	.7210	.7218	.7226	.7235
5.3	.7243	.7251	.7259	.7267	.7275	.7284	.7292	.7300	.7308	.7316
5.4	.7324	.7332	.7340	.7348	.7356	.7364	.7372	.7380	.7388	.7396
5.5	.7404	.7412	.7419	.7427	.7435	.7443	.7451	.7459	.7466	.7474
5.6	.7482	.7490	.7497	.7505	.7513	.7520	.7528	.7536	.7543	.7551
5.7	.7559	.7566	.7574	.7582	.7589	.7597	.7604	.7612	.7619	.7627
5.8	.7634	.7642	.7649	.7657	.7664	.7672	.7679	.7686	.7694	.7701
5.9	.7709	.7716	.7723	.7731	.7738	.7745	.7752	.7760	.7767	.7774
6.0	.7782	.7789	.7796	.7803	.7810	.7818	.7825	.7832	.7839	.7846
6.1	.7853	.7860	.7868	.7875	.7882	.7889	.7896	.7903	.7910	.7917
6.2	.7924	.7931	.7938	.7945	.7952	.7959	.7966	.7973	.7980	.7987
6.3	.7993	.8000	.8007	.8014	.8021	.8028	.8035	.8041	.8048	.8055
6.4	.8062	.8069	.8075	.8082	.8089	.8096	.8102	.8109	.8116	.8122
6.5	.8129	.8136	.8142	.8149	.8156	.8162	.8169	.8176	.8182	.8189
6.6	.8195	.8202	.8209	.8215	.8222	.8228	.8235	.8241	.8248	.8254
6.7	.8261	.8267	.8274	.8280	.8287	.8293	.8299	.8306	.8312	.8319
6.8	.8325	.8331	.8338	.8344	.8351	.8357	.8363	.8370	.8376	.8382
6.9	.8388	.8395	.8401	.8407	.8414	.8420	.8426	.8432	.8439	.8445
7.0	.8451	.8457	.8463	.8470	.8476	.8482	.8488	.8494	.8500	.8506
7.1	.8513	.8519	.8525	.8531	.8537	.8543	.8549	.8555	.8561	.8567
7.2	.8573	.8579	.8585	.8591	.8597	.8603	.8609	.8615	.8621	.8627
7.3	.8633	.8639	.8645	.8651	.8657	.8663	.8669	.8675	.8681	.8686
7.4	.8692	.8698	.8704	.8710	.8716	.8722	.8727	.8733	.8739	.8745
7.5	.8751	.8756	.8762	.8768	.8774	.8779	.8785	.8791	.8797	.8802
7.6	.8808	.8814	.8820	.8825	.8831	.8837	.8842	.8848	.8854	.8859
7.7	.8865	.8871	.8876	.8882	.8887	.8893	.8899	.8904	.8910	.8915
7.8	.8921	.8927	.8932	.8938	.8943	.8949	.8954	.8960	.8965	.8971
7.9	.8976	.8982	.8987	.8993	.8998	.9004	.9009	.9015	.9020	.9025
8.0	.9031	.9036	.9042	.9047	.9053	.9058	.9063	.9069	.9074	.9079
8.1	.9085	.9090	.9096	.9101	.9106	.9112	.9117	.9122	.9128	.9133
8.2	.9138	.9143	.9149	.9154	.9159	.9165	.9170	.9175	.9180	.9186
8.3	.9191	.9196	.9201	.9206	.9212	.9217	.9222	.9227	.9232	.9238
8.4	.9243	.9248	.9253	.9258	.9263	.9269	.9274	.9279	.9284	.9289
8.5	.9294	.9299	.9304	.9309	.9315	.9320	.9325	.9330	.9335	.9340
8.6	.9345	.9350	.9355	.9360	.9365	.9370	.9375	.9380	.9385	.9390
8.7	.9395	.9400	.9405	.9410	.9415	.9420	.9425	.9430	.9435	.9440
8.8	.9445	.9450	.9455	.9460	.9465	.9469	.9474	.9479	.9484	.9489
8.9	.9494	.9499	.9504	.9509	.9513	.9518	.9523	.9528	.9533	.9538
9.0	.9542	.9547	.9552	.9557	.9562	.9566	.9571	.9576	.9581	.9586
9.1	.9590	.9595	.9600	.9605	.9609	.9614	.9619	.9624	.9628	.9633
9.2	.9638	.9643	.9647	.9652	.9657	.9661	.9666	.9671	.9675	.9680
9.3	.9685	.9689	.9694	.9699	.9703	.9708	.9713	.9717	.9722	.9727
9.4	.9731	.9736	.9741	.9745	.9750	.9754	.9759	.9764	.9768	.9773
9.5	.9777	.9782	.9786	.9791	.9795	.9800	.9805	.9809	.9814	.9818
9.6	.9823	.9827	.9832	.9836	.9841	.9845	.9850	.9854	.9859	.9863
9.7	.9868	.9872	.9877	.9881	.9886	.9890	.9894	.9899	.9903	.9908
9.8	.9912	.9917	.9921	.9926	.9930	.9934	.9939	.9943	.9948	.9952
9.9	.9956	.9961	.9965	.9969	.9974	.9978	.9983	.9987	.9991	.9996

TABLE 2

Trigonometric Function Table

Degrees	Sin	Cos	Tan	Radians
0.0	0.00000	1.00000	0.00000	0.00000
0.2	0.00349	0.99999	0.00349	0.00349
0.4	0.00698	0.99998	0.00698	0.00698
0.6	0.01047	0.99995	0.01047	0.01047
0.8	0.01396	0.99990	0.01396	0.01396
1.0	0.01745	0.99985	0.01746	0.01745
1.2	0.02094	0.99978	0.02095	0.02094
1.4	0.02443	0.99970	0.02444	0.02443
1.6	0.02792	0.99961	0.02793	0.02793
1.8	0.03141	0.99951	0.03143	0.03142
2.0	0.03490	0.99939	0.03492	0.03491
2.2	0.03839	0.99926	0.03842	0.03840
2.4	0.04188	0.99912	0.04191	0.04189
2.6	0.04536	0.99897	0.04541	0.04538
2.8	0.04885	0.99881	0.04891	0.04887
3.0	0.05234	0.99863	0.05241	0.05236
3.2	0.05582	0.99844	0.05591	0.05585
3.4	0.05931	0.99824	0.05941	0.05934
3.6	0.06279	0.99803	0.06291	0.06283
3.8	0.06627	0.99780	0.06642	0.06632
4.0	0.06976	0.99756	0.06993	0.06981
4.2	0.07324	0.99731	0.07344	0.07330
4.4	0.07672	0.99705	0.07695	0.07679
4.6	0.08020	0.99678	0.08046	0.08029
4.8	0.08368	0.99649	0.08397	0.08378
5.0	0.08716	0.99619	0.08749	0.08727
5.2	0.09063	0.99588	0.09101	0.09076
5.4	0.09411	0.99556	0.09453	0.09425
5.6	0.09758	0.99523	0.09805	0.09774
5.8	0.10106	0.99488	0.10158	0.10123
6.0	0.10453	0.99452	0.10510	0.10472
6.2	0.10800	0.99415	0.10863	0.10821
6.4	0.11147	0.99377	0.11217	0.11170
6.6	0.11494	0.99337	0.11570	0.11519
6.8	0.11840	0.99297	0.11924	0.11868
7.0	0.12187	0.99255	0.12278	0.12217
7.2	0.12533	0.99211	0.12633	0.12566
7.4	0.12880	0.99167	0.12988	0.12915
7.6	0.13226	0.99122	0.13343	0.13264
7.8	0.13572	0.99075	0.13698	0.13614
8.0	0.13917	0.99027	0.14054	0.13963
8.2	0.14263	0.98978	0.14410	0.14312
8.4	0.14608	0.98927	0.14767	0.14661
8.6	0.14954	0.98876	0.15124	0.15010
8.8	0.15299	0.98823	0.15481	0.15359
9.0	0.15643	0.98769	0.15838	0.15708
9.2	0.15988	0.98714	0.16196	0.16057
9.4	0.16333	0.98657	0.16555	0.16406
9.6	0.16677	0.98600	0.16914	0.16755
9.8	0.17021	0.98541	0.17273	0.17104

Degrees	Sin	Cos	Tan	Radians
10.0	0.17365	0.98481	0.17633	0.17453
10.2	0.17708	0.98420	0.17993	0.17802
10.4	0.18052	0.98357	0.18353	0.18151
10.6	0.18395	0.98294	0.18714	0.18500
10.8	0.18738	0.98229	0.19076	0.18850
11.0	0.19081	0.98163	0.19438	0.19199
11.2	0.19423	0.98096	0.19801	0.19548
11.4	0.19766	0.98027	0.20164	0.19897
11.6	0.20108	0.97958	0.20527	0.20246
11.8	0.20450	0.97887	0.20891	0.20595
12.0	0.20791	0.97815	0.21256	0.20944
12.2	0.21132	0.97742	0.21621	0.21293
12.4	0.21474	0.97667	0.21986	0.21642
12.6	0.21814	0.97592	0.22353	0.21991
12.8	0.22155	0.97515	0.22719	0.22340
13.0	0.22495	0.97437	0.23087	0.22689
13.2	0.22835	0.97358	0.23455	0.23038
13.4	0.23175	0.97278	0.23823	0.23387
13.6	0.23514	0.97196	0.24193	0.23736
13.8	0.23853	0.97113	0.24562	0.24086
14.0	0.24192	0.97030	0.24933	0.24435
14.2	0.24531	0.96945	0.25304	0.24784
14.4	0.24869	0.96858	0.25676	0.25133
14.6	0.25207	0.96771	0.26048	0.25482
14.8	0.25545	0.96682	0.26421	0.25831
15.0	0.25882	0.96593	0.26795	0.26180
15.2	0.26219	0.96502	0.27169	0.26529
15.4	0.26556	0.96410	0.27545	0.26878
15.6	0.26892	0.96316	0.27920	0.27227
15.8	0.27228	0.96222	0.28297	0.27576
16.0	0.27564	0.96126	0.28675	0.27925
16.2	0.27899	0.96029	0.29053	0.28274
16.4	0.28234	0.95931	0.29432	0.28623
16.6	0.28569	0.95832	0.29811	0.28972
16.8	0.28903	0.95732	0.30192	0.29322
17.0	0.29237	0.95631	0.30573	0.29671
17.2	0.29571	0.95528	0.30955	0.30020
17.4	0.29904	0.95424	0.31338	0.30369
17.6	0.30237	0.95319	0.31722	0.30718
17.8	0.30570	0.95213	0.32106	0.31067
18.0	0.30902	0.95106	0.32492	0.31416
18.2	0.31233	0.94997	0.32878	0.31765
18.4	0.31565	0.94888	0.33266	0.32114
18.6	0.31896	0.94777	0.33654	0.32463
18.8	0.32227	0.94665	0.34033	0.32812
19.0	0.32557	0.94552	0.34433	0.33161
19.2	0.32887	0.94438	0.34824	0.33510
19.4	0.33216	0.94322	0.35216	0.33859
19.6	0.33545	0.94206	0.35608	0.34208
19.8	0.33874	0.94088	0.36002	0.34557
20.0	0.34202	0.93969	0.36397	0.34907
20.2	0.34530	0.93849	0.36793	0.35256
20.4	0.34857	0.93728	0.37190	0.35605
20.6	0.35184	0.93606	0.37587	0.35954
20.8	0.35511	0.93483	0.37986	0.36303

Degrees	Sin	Cos	Tan	Radians
21.0	0.35837	0.93358	0.38386	0.36652
21.2	0.36162	0.93232	0.38787	0.37001
21.4	0.36488	0.93106	0.39190	0.37350
21.6	0.36812	0.92978	0.39593	0.37699
21.8	0.37137	0.92849	0.39997	0.38048
22.0	0.37461	0.92718	0.40403	0.38397
22.2	0.37784	0.92587	0.40809	0.38746
22.4	0.38107	0.92455	0.41217	0.39095
22.6	0.38430	0.92321	0.41626	0.39444
22.8	0.38752	0.92186	0.42036	0.39793
23.0	0.39073	0.92051	0.42447	0.40143
23.2	0.39394	0.91914	0.42860	0.40492
23.4	0.39715	0.91775	0.43274	0.40841
23.6	0.40035	0.91636	0.43689	0.41190
23.8	0.40354	0.91496	0.44105	0.41539
24.0	0.40674	0.91355	0.44523	0.41888
24.2	0.40992	0.91212	0.44942	0.42237
24.4	0.41310	0.91068	0.45362	0.42586
24.6	0.41628	0.90924	0.45784	0.42935
24.8	0.41945	0.90778	0.46206	0.43284
25.0	0.42262	0.90631	0.46631	0.43633
25.2	0.42578	0.90483	0.47056	0.43982
25.4	0.42893	0.90334	0.47483	0.44331
25.6	0.43209	0.90183	0.47912	0.44680
25.8	0.43523	0.90032	0.48342	0.45029
26.0	0.43837	0.89879	0.48773	0.45379
26.2	0.44151	0.89726	0.49206	0.45728
26.4	0.44463	0.89571	0.49640	0.46077
26.6	0.44776	0.89415	0.50076	0.46426
26.8	0.45088	0.89259	0.50514	0.46775
27.0	0.45399	0.89101	0.50952	0.47124
27.2	0.45710	0.88942	0.51393	0.47473
27.4	0.46020	0.88782	0.51835	0.47822
27.6	0.46330	0.88620	0.52279	0.48171
27.8	0.46639	0.88458	0.52724	0.48520
28.0	0.46947	0.88295	0.53171	0.48869
28.2	0.47255	0.88130	0.53619	0.49218
28.4	0.47562	0.87965	0.54070	0.49567
28.6	0.47869	0.87798	0.54522	0.49916
28.8	0.48175	0.87631	0.54975	0.50265
29.0	0.48481	0.87462	0.55431	0.50615
29.2	0.48786	0.87292	0.55888	0.50964
29.4	0.49090	0.87121	0.56347	0.51313
29.6	0.49394	0.86950	0.56808	0.51662
29.8	0.49697	0.86777	0.57270	0.52011
30.0	0.50000	0.86603	0.57735	0.52360
30.2	0.50302	0.86428	0.58201	0.52709
30.4	0.50603	0.86251	0.58670	0.53058
30.6	0.50904	0.86074	0.59140	0.53407
30.8	0.51204	0.85896	0.59612	0.53756
31.0	0.51504	0.85717	0.60086	0.54105
31.2	0.51803	0.85536	0.60562	0.54454
31.4	0.52101	0.85355	0.61040	0.54803
31.6	0.52399	0.85173	0.61520	0.55152
31.8	0.52696	0.84989	0.62003	0.55501

Degrees	Sin	Cos	Tan	Radians
32.0	0.52992	0.84805	0.62487	0.55850
32.2	0.53288	0.84619	0.62973	0.56200
32.4	0.53583	0.84433	0.63462	0.56549
32.6	0.53877	0.84245	0.63953	0.56898
32.8	0.54171	0.84057	0.64446	0.57247
33.0	0.54464	0.83867	0.64941	0.57596
33.2	0.54756	0.83676	0.65438	0.57945
33.4	0.55048	0.83485	0.65938	0.58294
33.6	0.55339	0.83292	0.66440	0.58643
33.8	0.55630	0.83098	0.66944	0.58992
34.0	0.55919	0.82904	0.67451	0.59341
34.2	0.56208	0.82708	0.67960	0.59690
34.4	0.56497	0.82511	0.68471	0.60039
34.6	0.56784	0.82314	0.68985	0.60388
34.8	0.57071	0.82115	0.69502	0.60737
35.0	0.57358	0.81915	0.70021	0.61086
35.2	0.57643	0.81715	0.70542	0.61436
35.4	0.57928	0.81513	0.71066	0.61785
35.6	0.58212	0.81310	0.71593	0.62134
35.8	0.58496	0.81106	0.72122	0.62483
36.0	0.58778	0.80902	0.72654	0.62832
36.2	0.59061	0.80696	0.73189	0.63181
36.4	0.59342	0.80489	0.73726	0.63530
36.6	0.59622	0.80282	0.74266	0.63879
36.8	0.59902	0.80073	0.74809	0.64228
37.0	0.60181	0.79864	0.75355	0.64577
37.2	0.60460	0.79653	0.75904	0.64926
37.4	0.60738	0.79442	0.76456	0.65275
37.6	0.61014	0.79229	0.77010	0.65624
37.8	0.61291	0.79016	0.77568	0.65973
38.0	0.61566	0.78801	0.78128	0.66322
38.2	0.61841	0.78586	0.78692	0.66672
38.4	0.62115	0.78369	0.79259	0.67021
38.6	0.62388	0.78152	0.79829	0.67370
38.8	0.62660	0.77934	0.80402	0.67719
39.0	0.62932	0.77715	0.80978	0.68068
39.2	0.63203	0.77495	0.81558	0.68417
39.4	0.63473	0.77273	0.82141	0.68766
39.6	0.63742	0.77051	0.82727	0.69115
39.8	0.64011	0.76828	0.83317	0.69464
40.0	0.64279	0.76604	0.83910	0.69813
40.2	0.64546	0.76380	0.84506	0.70162
40.4	0.64812	0.76154	0.85107	0.70511
40.6	0.65077	0.75927	0.85710	0.70860
40.8	0.65342	0.75700	0.86318	0.71209
41.0	0.65606	0.75471	0.86929	0.71558
41.2	0.65869	0.75242	0.87543	0.71908
41.4	0.66131	0.75011	0.88162	0.72257
41.6	0.66393	0.74780	0.88784	0.72606
41.8	0.66653	0.74548	0.89410	0.72955
42.0	0.66913	0.74315	0.90040	0.73304
42.2	0.67172	0.74081	0.90674	0.73653
42.4	0.67430	0.73846	0.91312	0.74002
42.6	0.67688	0.73610	0.91955	0.74351
42.8	0.67944	0.73373	0.92601	0.74700

Degrees	Sin	Cos	Tan	Radians
43.0	0.68200	0.73135	0.93251	0.75049
43.2	0.68455	0.72897	0.93906	0.75398
43.4	0.68709	0.72658	0.94565	0.75747
43.6	0.68962	0.72417	0.95229	0.76096
43.8	0.69214	0.72176	0.95896	0.76445
44.0	0.69466	0.71934	0.96569	0.76794
44.2	0.69716	0.71691	0.97246	0.77144
44.4	0.69966	0.71447	0.97927	0.77493
44.6	0.70215	0.71203	0.98613	0.77842
44.8	0.70463	0.70957	0.99304	0.78191
45.0	0.70711	0.70711	1.00000	0.78540
45.2	0.70957	0.70463	1.00700	0.78889
45.4	0.71203	0.70215	1.01406	0.79238
45.6	0.71447	0.69966	1.02116	0.79587
45.8	0.71691	0.69717	1.02832	0.79936
46.0	0.71934	0.69466	1.03553	0.80285
46.2	0.72176	0.69214	1.04279	0.80634
46.4	0.72417	0.68962	1.05010	0.80983
46.6	0.72657	0.68709	1.05747	0.81332
46.8	0.72897	0.68455	1.06489	0.81681
47.0	0.73135	0.68200	1.07237	0.82030
47.2	0.73373	0.67944	1.07990	0.82379
47.4	0.73610	0.67688	1.08749	0.82729
47.6	0.73845	0.67430	1.09514	0.83078
47.8	0.74080	0.67172	1.10284	0.83427
48.0	0.74314	0.66913	1.11061	0.83776
48.2	0.74548	0.66653	1.11844	0.84125
48.4	0.74780	0.66393	1.12633	0.84474
48.6	0.75011	0.66131	1.13428	0.84823
48.8	0.75241	0.65869	1.14229	0.85172
49.0	0.75471	0.65606	1.15037	0.85521
49.2	0.75699	0.65342	1.15851	0.85870
49.4	0.75927	0.65077	1.16672	0.86219
49.6	0.76154	0.64812	1.17499	0.86568
49.8	0.76380	0.64546	1.18334	0.86917
50.0	0.76604	0.64279	1.19175	0.87266
50.2	0.76828	0.64011	1.20024	0.87615
50.4	0.77051	0.63742	1.20879	0.87965
50.6	0.77273	0.63473	1.21742	0.88314
50.8	0.77494	0.63203	1.22612	0.88663
51.0	0.77715	0.62932	1.23490	0.89012
51.2	0.77934	0.62660	1.24375	0.89361
51.4	0.78152	0.62388	1.25268	0.89710
51.6	0.78369	0.62115	1.26168	0.90059
51.8	0.78586	0.61841	1.27077	0.90408
52.0	0.78801	0.61566	1.27994	0.90757
52.2	0.79015	0.61291	1.28919	0.91106
52.4	0.79229	0.61015	1.29852	0.91455
52.6	0.79441	0.60738	1.30794	0.91804
52.8	0.79653	0.60460	1.31745	0.92153
53.0	0.79864	0.60182	1.32074	0.92502
53.2	0.80073	0.59902	1.33673	0.92851
53.4	0.80282	0.59623	1.34650	0.93201
53.6	0.80489	0.59342	1.35636	0.93550
53.8	0.80696	0.59061	1.36632	0.93899

Degrees	Sin	Cos	Tan	Radians
54.0	0.80902	0.58779	1.37638	0.94248
54.2	0.81106	0.58496	1.38653	0.94597
54.4	0.81310	0.58212	1.39678	0.94946
54.6	0.81513	0.57928	1.40713	0.95295
54.8	0.81714	0.57643	1.41759	0.95644
55.0	0.81915	0.57358	1.42815	0.95993
55.2	0.82115	0.57071	1.43881	0.96342
55.4	0.82314	0.56784	1.44958	0.96691
55.6	0.82511	0.56497	1.46046	0.97040
55.8	0.82708	0.56208	1.47145	0.97389
56.0	0.82904	0.55919	1.48256	0.97738
56.2	0.83098	0.55630	1.49378	0.98087
56.4	0.83292	0.55339	1.50512	0.98436
56.6	0.83485	0.55048	1.51658	0.98786
56.8	0.83676	0.54756	1.52816	0.99135
57.0	0.83867	0.54464	1.53986	0.99484
57.2	0.84057	0.54171	1.55169	0.99833
57.4	0.84245	0.53877	1.56365	1.00182
57.6	0.84433	0.53583	1.57574	1.00531
57.8	0.84619	0.53288	1.58797	1.00880
58.0	0.84805	0.52992	1.60033	1.01229
58.2	0.84989	0.52696	1.61283	1.01578
58.4	0.85173	0.52399	1.62547	1.01927
58.6	0.85355	0.52101	1.63826	1.02276
58.8	0.85536	0.51803	1.65119	1.02625
59.0	0.85717	0.51504	1.66428	1.02974
59.2	0.85896	0.51204	1.67751	1.03323
59.4	0.86074	0.50904	1.69090	1.03672
59.6	0.86251	0.50603	1.70446	1.04022
59.8	0.86427	0.50302	1.71817	1.04371
60.0	0.86602	0.50000	1.73205	1.04720
60.2	0.86777	0.49697	1.74610	1.05069
60.4	0.86949	0.49394	1.76032	1.05418
60.6	0.87121	0.49090	1.77471	1.05767
60.8	0.87292	0.48786	1.78929	1.06116
61.0	0.87462	0.48481	1.80404	1.06465
61.2	0.87631	0.48175	1.81899	1.06814
61.4	0.87798	0.47869	1.83413	1.07163
61.6	0.87965	0.47563	1.84946	1.07512
61.8	0.88130	0.47255	1.86499	1.07861
62.0	0.88295	0.46947	1.88072	1.08210
62.2	0.88458	0.46639	1.89666	1.08559
62.4	0.88620	0.46330	1.91282	1.08908
62.6	0.88781	0.46020	1.92919	1.09258
62.8	0.88942	0.45710	1.94578	1.09607
63.0	0.89101	0.45399	1.96261	1.09956
63.2	0.89259	0.45088	1.97966	1.10305
63.4	0.89415	0.44776	1.99695	1.10654
63.6	0.89571	0.44464	2.01448	1.11003
63.8	0.89726	0.44151	2.03226	1.11352
64.0	0.89879	0.43837	2.05030	1.11701
64.2	0.90032	0.43523	2.06859	1.12050
64.4	0.90183	0.43209	2.08716	1.12399
64.6	0.90333	0.42894	2.10599	1.12748
64.8	0.90483	0.42578	2.12510	1.13097

Degrees	Sin	Cos	Tan	Radians
65.0	0.90631	0.42262	2.14450	1.13446
65.2	0.90778	0.41945	2.16419	1.13795
65.4	0.90924	0.41628	2.18418	1.14144
65.6	0.91068	0.41311	2.20448	1.14494
65.8	0.91212	0.40992	2.22510	1.14843
66.0	0.91355	0.40674	2.24603	1.15192
66.2	0.91496	0.40355	2.26730	1.15541
66.4	0.91636	0.40035	2.28890	1.15890
66.6	0.91775	0.39715	2.31086	1.16239
66.8	0.91914	0.39394	2.33317	1.16588
67.0	0.92050	0.39073	2.35585	1.16937
67.2	0.92186	0.38752	2.37890	1.17286
67.4	0.92321	0.38430	2.40234	1.17635
67.6	0.92455	0.38107	2.42617	1.17984
67.8	0.92587	0.37784	2.45042	1.18333
68.0	0.92718	0.37461	2.47508	1.18682
68.2	0.92849	0.37137	2.50017	1.19031
68.4	0.92978	0.36813	2.52570	1.19380
68.6	0.93106	0.36488	2.55169	1.19729
68.8	0.93232	0.36163	2.57815	1.20079
69.0	0.93358	0.35837	2.60508	1.20428
69.2	0.93483	0.35511	2.63251	1.20777
69.4	0.93606	0.35184	2.66045	1.21126
69.6	0.93728	0.34857	2.68891	1.21475
69.8	0.93849	0.34530	2.71791	1.21824
70.0	0.93969	0.34202	2.74747	1.22173
70.2	0.94088	0.33874	2.77760	1.22522
70.4	0.94206	0.33545	2.80832	1.22871
70.6	0.94322	0.33216	2.83964	1.23220
70.8	0.94438	0.32887	2.87160	1.23569
71.0	0.94552	0.32557	2.90420	1.23918
71.2	0.94665	0.32227	2.93747	1.24267
71.4	0.94777	0.31896	2.97143	1.24616
71.6	0.94888	0.31565	3.00610	1.24965
71.8	0.94997	0.31234	3.04151	1.25315
72.0	0.95106	0.30902	3.07767	1.25664
72.2	0.95213	0.30570	3.11462	1.26013
72.4	0.95319	0.30237	3.15239	1.26362
72.6	0.95424	0.29904	3.19099	1.26711
72.8	0.95528	0.29571	3.23047	1.27060
73.0	0.95630	0.29237	3.27084	1.27409
73.2	0.95732	0.28903	3.31215	1.27758
73.4	0.95832	0.28569	3.35442	1.28107
73.6	0.95931	0.28234	3.39769	1.28456
73.8	0.96029	0.27899	3.44201	1.28805
74.0	0.96126	0.27564	3.48740	1.29154
74.2	0.96222	0.27228	3.53391	1.29503
74.4	0.96316	0.26892	3.58158	1.29852
74.6	0.96410	0.26556	3.63046	1.30201
74.8	0.96502	0.26219	3.68059	1.30551
75.0	0.96593	0.25882	3.73203	1.30900
75.2	0.96682	0.25545	3.78483	1.31249
75.4	0.96771	0.25207	3.83904	1.31598
75.6	0.96858	0.24869	3.89473	1.31947
75.8	0.96945	0.24531	3.95194	1.32296

Degrees	Sin	Cos	Tan	Radians
76.0	0.97030	0.24192	4.01076	1.32645
76.2	0.97113	0.23853	4.07125	1.32994
76.4	0.97196	0.23514	4.13348	1.33343
76.6	0.97278	0.23175	4.19754	1.33692
76.8	0.97358	0.22835	4.26350	1.34041
77.0	0.97437	0.22495	4.33145	1.34390
77.2	0.97515	0.22155	4.40149	1.34739
77.4	0.97592	0.21814	4.47372	1.35088
77.6	0.97667	0.21474	4.54823	1.35437
77.8	0.97742	0.21133	4.62516	1.35787
78.0	0.97815	0.20791	4.70460	1.36136
78.2	0.97887	0.20450	4.78670	1.36485
78.4	0.97958	0.20108	4.87159	1.36834
78.6	0.98027	0.19766	4.95942	1.37183
78.8	0.98096	0.19424	5.05034	1.37532
79.0	0.98163	0.19081	5.14452	1.37881
79.2	0.98229	0.18738	5.24215	1.38230
79.4	0.98294	0.18395	5.34342	1.38579
79.6	0.98357	0.18052	5.44853	1.38928
79.8	0.98420	0.17709	5.55773	1.39277
80.0	0.98481	0.17365	5.67124	1.39626
80.2	0.98541	0.17021	5.78935	1.39975
80.4	0.98600	0.16677	5.91231	1.40324
80.6	0.98657	0.16333	6.04046	1.40673
80.8	0.98714	0.15988	6.17414	1.41023
81.0	0.98769	0.15644	6.31370	1.41372
81.2	0.98823	0.15299	6.45956	1.41721
81.4	0.98876	0.14954	6.61213	1.42070
81.6	0.98927	0.14608	6.77193	1.42419
81.8	0.98978	0.14263	6.93946	1.42768
82.0	0.99027	0.13917	7.11531	1.43117
82.2	0.99075	0.13572	7.30010	1.43466
82.4	0.99122	0.13226	7.49458	1.43815
82.6	0.99167	0.12880	7.69950	1.44164
82.8	0.99211	0.12533	7.91574	1.44513
83.0	0.99255	0.12187	8.14426	1.44862
83.2	0.99297	0.11841	8.38617	1.45211
83.4	0.99337	0.11494	8.64266	1.45560
83.6	0.99377	0.11147	8.91509	1.45909
83.8	0.99415	0.10800	9.20506	1.46258
84.0	0.99452	0.10453	9.51424	1.46608
84.2	0.99488	0.10106	9.84469	1.46957
84.4	0.99523	0.09758	10.19860	1.47306
84.6	0.99556	0.09411	10.57880	1.47655
84.8	0.99588	0.09063	10.98800	1.48004
85.0	0.99619	0.08716	11.42990	1.48353
85.2	0.99649	0.08368	11.90850	1.48702
85.4	0.99678	0.08020	12.42860	1.49051
85.6	0.99705	0.07672	12.99590	1.49400
85.8	0.99731	0.07324	13.61710	1.49749
86.0	0.99756	0.06976	14.30040	1.50098
86.2	0.99780	0.06628	15.05540	1.50447
86.4	0.99803	0.06279	15.89420	1.50796
86.6	0.99824	0.05931	16.83150	1.51145
86.8	0.99844	0.05582	17.88590	1.51494

Degrees	Sin	Cos	Tan	Radians
87.0	0.99863	0.05234	19.08060	1.51844
87.2	0.99881	0.04885	20.44590	1.52193
87.4	0.99897	0.04536	22.02100	1.52542
87.6	0.99912	0.04188	23.85860	1.52891
87.8	0.99926	0.03839	26.02980	1.53240
88.0	0.99939	0.03490	28.63530	1.53589
88.2	0.99951	0.03141	31.81900	1.53938
88.4	0.99961	0.02792	35.79910	1.54287
88.6	0.99970	0.02443	40.91510	1.54636
88.8	0.99978	0.02094	47.73610	1.54985
89.0	0.99985	0.01745	57.28550	1.55334
89.2	0.99990	0.01396	71.60780	1.55683
89.4	0.99995	0.01047	95.47760	1.56032
89.6	0.99998	0.00698	143.20900	1.56381
89.8	0.99999	0.00349	286.37600	1.56730

TABLE 3

The Standard Normal Distribution

If Z has a standard normal distribution, the table gives the value of $\Pr(Z < z)$.

z	$\Pr(Z < z)$	z	$\Pr(Z < z)$	z	$\Pr(Z < z)$	z	$\Pr(Z < z)$
0.01	.5040	0.29	.6141	0.57	.7157	0.85	.8023
0.02	.5080	0.30	.6179	0.58	.7190	0.86	.8051
0.03	.5120	0.31	.6217	0.59	.7224	0.87	.8079
0.04	.5160	0.32	.6255	0.60	.7257	0.88	.8106
0.05	.5199	0.33	.6293	0.61	.7291	0.89	.8133
0.06	.5239	0.34	.6331	0.62	.7324	0.90	.8159
0.07	.5279	0.35	.6368	0.63	.7357	0.91	.8186
0.08	.5319	0.36	.6406	0.64	.7389	0.92	.8212
0.09	.5359	0.37	.6443	0.65	.7422	0.93	.8238
0.10	.5398	0.38	.6480	0.66	.7454	0.94	.8264
0.11	.5438	0.39	.6517	0.67	.7486	0.95	.8289
0.12	.5478	0.40	.6554	0.68	.7517	0.96	.8315
0.13	.5517	0.41	.6591	0.69	.7549	0.97	.8340
0.14	.5557	0.42	.6628	0.70	.7580	0.98	.8365
0.15	.5596	0.43	.6664	0.71	.7611	0.99	.8389
0.16	.5636	0.44	.6700	0.72	.7642	1.00	.8413
0.17	.5675	0.45	.6736	0.73	.7673	1.01	.8438
0.18	.5714	0.46	.6772	0.74	.7704	1.02	.8461
0.19	.5753	0.47	.6808	0.75	.7734	1.03	.8485
0.20	.5793	0.48	.6844	0.76	.7764	1.04	.8508
0.21	.5832	0.49	.6879	0.77	.7794	1.05	.8531
0.22	.5871	0.50	.6915	0.78	.7823	1.06	.8554
0.23	.5910	0.51	.6950	0.79	.7852	1.07	.8577
0.24	.5948	0.52	.6985	0.80	.7881	1.08	.8599
0.25	.5987	0.53	.7019	0.81	.7910	1.09	.8621
0.26	.6026	0.54	.7054	0.82	.7939	1.10	.8643
0.27	.6064	0.55	.7088	0.83	.7967	1.11	.8665
0.28	.6103	0.56	.7123	0.84	.7995	1.12	.8686

z	$\Pr(Z < z)$	z	$\Pr(Z < z)$	z	$\Pr(Z < z)$	z	$\Pr(Z < z)$
1.13	.8708	1.44	.9251	1.75	.9599	2.06	.9803
1.14	.8729	1.45	.9265	1.76	.9608	2.07	.9808
1.15	.8749	1.46	.9279	1.77	.9616	2.08	.9812
1.16	.8770	1.47	.9292	1.78	.9625	2.09	.9817
1.17	.8790	1.48	.9306	1.79	.9633	2.10	.9821
1.18	.8810	1.49	.9319	1.80	.9641	2.11	.9826
1.19	.8830	1.50	.9332	1.81	.9649	2.12	.9830
1.20	.8849	1.51	.9345	1.82	.9656	2.13	.9834
1.21	.8869	1.52	.9357	1.83	.9664	2.14	.9838
1.22	.8888	1.53	.9370	1.84	.9671	2.15	.9842
1.23	.8907	1.54	.9382	1.85	.9678	2.16	.9846
1.24	.8925	1.55	.9394	1.86	.9686	2.17	.9850
1.25	.8944	1.56	.9406	1.87	.9693	2.18	.9854
1.26	.8962	1.57	.9418	1.88	.9699	2.19	.9857
1.27	.8980	1.58	.9429	1.89	.9706	2.20	.9861
1.28	.8997	1.59	.9441	1.90	.9713	2.25	.9878
1.29	.9015	1.60	.9452	1.91	.9719	2.30	.9893
1.30	.9032	1.61	.9463	1.92	.9726	2.35	.9906
1.31	.9049	1.62	.9474	1.93	.9732	2.40	.9918
1.32	.9066	1.63	.9484	1.94	.9738	2.50	.9938
1.33	.9082	1.64	.9495	1.95	.9744	2.60	.9953
1.34	.9099	1.65	.9505	1.96	.9750	2.70	.9965
1.35	.9115	1.66	.9515	1.97	.9756	2.80	.9974
1.36	.9131	1.67	.9525	1.98	.9761	2.90	.9981
1.37	.9147	1.68	.9535	1.99	.9767	3.00	.9987
1.38	.9162	1.69	.9545	2.00	.9773	3.10	.9990
1.39	.9177	1.70	.9554	2.01	.9778	3.20	.9993
1.40	.9192	1.71	.9564	2.02	.9783	3.30	.9995
1.41	.9207	1.72	.9573	2.03	.9788	3.40	.9997
1.42	.9222	1.73	.9582	2.04	.9793	3.50	.9998
1.43	.9236	1.74	.9591	2.05	.9798		

TABLE 4

The Standard Normal Distribution

z	$\Pr(-z < Z < z)$	z	$\Pr(-z < Z < z)$
0.10	.0796	1.30	.8064
0.20	.1586	1.40	.8384
0.30	.2358	1.50	.8664
0.40	.3108	1.60	.8904
0.50	.3830	1.70	.9108
0.60	.4514	1.80	.9282
0.70	.5160	1.90	.9426
0.80	.5762	1.96	.9500
0.90	.6318	2.00	.9546
1.00	.6826	2.50	.9876
1.10	.7286	3.00	.9974

TABLE 5

The Chi-square Cumulative Distribution Function

If X has a chi-square distribution with n degrees of freedom, the table gives the value x such that $\Pr(X < x) = p$. For example, if X has a chi-square distribution with 10 degrees of freedom, there is a probability of .95 that X will be less than 18.3.

n	$p=.005$	$p=.01$	$p=.025$	$p=.05$	$p=.50$	$p=.75$	$p=.90$	$p=.95$	$p=.975$	$p=.99$
1	.000	.000	.001	.004	.45	1.32	2.71	3.84	5.02	6.64
2	.01	.02	.05	.10	1.38	2.77	4.60	5.99	7.37	9.21
3	.07	.11	.22	.35	2.36	4.10	6.24	7.80	9.33	11.31
4	.20	.29	.48	.71	3.35	5.38	7.77	9.48	11.14	13.27
5	.41	.55	.93	1.14	4.35	6.62	9.23	11.07	12.83	15.08
6	.67	.87	1.24	1.63	5.34	7.84	10.64	12.59	14.44	16.81
7	.98	1.24	1.69	2.17	6.35	9.04	12.02	14.07	16.01	18.48
8	1.34	1.65	2.18	2.73	7.34	10.22	13.36	15.51	17.54	20.09
9	1.73	2.09	2.70	3.33	8.34	11.39	14.68	16.92	19.02	21.67
10	2.16	2.56	3.25	3.94	9.3	12.5	15.9	18.3	20.5	23.2
11	2.60	3.05	3.82	4.57	10.3	13.7	17.3	19.7	21.9	24.7
12	3.07	3.57	4.40	5.23	11.3	14.8	18.6	21.0	23.3	26.2
13	3.56	4.11	5.01	5.89	12.3	16.0	19.8	22.4	24.7	27.7
14	4.08	4.66	5.63	6.57	13.3	17.1	21.1	23.7	26.1	29.1
15	4.60	5.23	6.26	7.26	14.3	18.2	22.3	25.0	27.5	30.6
16	5.14	5.81	6.91	7.96	15.3	19.4	23.5	26.3	28.8	32.0
17	5.70	6.41	7.56	8.67	16.3	20.5	24.8	27.6	30.2	33.4
18	6.26	7.02	8.23	9.39	17.3	21.6	26.0	28.9	31.5	34.8
19	6.85	7.63	8.91	10.12	18.3	22.7	27.2	30.1	32.9	36.2
20	7.43	8.26	9.59	10.85	19.3	23.8	28.4	31.4	34.2	37.6
21	8.03	8.90	10.28	11.59	20.3	24.9	29.6	32.7	35.5	38.9
22	8.64	9.54	10.98	12.34	21.3	26.0	30.8	33.9	36.8	40.3
23	9.26	10.19	11.69	13.09	22.3	27.1	32.0	35.2	38.1	41.6
24	9.89	10.86	12.40	13.85	23.3	28.2	33.2	36.4	39.4	43.0
25	10.52	11.52	13.12	14.61	24.3	29.3	34.4	37.7	40.7	44.3
30	13.79	14.95	16.79	18.49	29.3	34.8	40.3	43.8	47.0	50.9
40	20.70	22.16	24.43	26.51	39.3	45.6	51.8	55.7	59.3	63.7
50	27.99	29.70	32.26	34.76	49.3	56.3	63.2	67.5	71.4	76.2
60	35.53	37.48	40.48	43.19	59.3	67.0	74.4	79.1	83.3	88.4
70	43.27	45.44	48.76	51.74	69.3	77.6	85.5	90.5	95.0	100.4
80	51.18	53.54	57.15	60.38	79.3	88.1	96.6	101.9	106.6	112.3
90	59.19	61.74	65.65	69.12	89.3	98.7	107.6	113.2	118.1	124.1

TABLE 6

The *t*-Distribution

If *X* has a *t*-distribution with *n* degrees of freedom, the table gives the value of *x* such that $Pr(X < x) = p$. For example, if *X* has a *t*-distribution with 15 degrees of freedom, there is a 95 percent chance *X* will be less than 1.753.

n	p = .750	p = .900	p = .950	p = .975	p = .990	p = .995
1	1.000	3.078	6.314	12.706	31.821	63.657
2	0.817	1.886	2.920	4.303	6.965	9.925
3	0.765	1.638	2.353	3.182	4.541	5.841
4	0.741	1.533	2.132	2.776	3.747	4.604
5	0.727	1.476	2.015	2.571	3.365	4.032
6	0.718	1.440	1.943	2.447	3.143	3.707
7	0.711	1.415	1.895	2.365	3.000	3.499
8	0.706	1.397	1.860	2.306	2.896	3.355
9	0.703	1.383	1.833	2.262	2.821	3.250
10	0.700	1.372	1.812	2.228	2.764	3.169
11	0.697	1.363	1.796	2.201	2.718	3.106
12	0.695	1.356	1.782	2.179	2.681	3.055
13	0.694	1.350	1.771	2.160	2.650	3.012
14	0.692	1.345	1.761	2.145	2.600	2.977
15	0.691	1.341	1.753	2.131	2.600	2.947
16	0.690	1.337	1.746	2.120	2.584	2.921
17	0.689	1.333	1.740	2.110	2.567	2.898
18	0.688	1.330	1.734	2.101	2.552	2.878
19	0.688	1.328	1.729	2.093	2.539	2.861
20	0.687	1.325	1.725	2.086	2.528	2.845
21	0.686	1.323	1.721	2.080	2.518	2.831
22	0.686	1.321	1.717	2.074	2.508	2.819
23	0.685	1.319	1.714	2.069	2.500	2.807
24	0.685	1.318	1.711	2.064	2.492	2.797
25	0.684	1.316	1.708	2.060	2.485	2.787
26	0.684	1.315	1.706	2.056	2.479	2.779
27	0.684	1.314	1.703	2.052	2.473	2.771
28	0.683	1.313	1.701	2.048	2.467	2.763
29	0.683	1.311	1.699	2.045	2.462	2.756
30	0.683	1.310	1.697	2.042	2.457	2.750
35	0.682	1.306	1.690	2.030	2.438	2.724
40	0.681	1.303	1.684	2.021	2.423	2.704
50	0.679	1.299	1.676	2.009	2.400	2.678
60	0.679	1.296	1.671	2.000	2.400	2.660
100	0.677	1.290	1.660	1.984	2.364	2.626
120	0.677	1.289	1.658	1.980	2.358	2.617

TABLE 7

The *t*-Distribution

If X has a t-distribution with n degrees of freedom, the table gives the value of x such that $\Pr(-x < X < x) = p$.

n	$p = .95$	$p = .99$	n	$p = .95$	$p = .99$
1	12.706	63.657	20	2.086	2.845
2	4.303	9.925	21	2.080	2.831
3	3.182	5.841	22	2.074	2.819
4	2.776	4.604	23	2.069	2.807
5	2.571	4.032	24	2.064	2.797
6	2.447	3.707	25	2.060	2.787
7	2.365	3.499	26	2.056	2.779
8	2.306	3.355	27	2.052	2.771
9	2.262	3.250	28	2.048	2.763
10	2.228	3.169	29	2.045	2.756
11	2.201	3.106	30	2.042	2.750
12	2.179	3.055	35	2.030	2.724
13	2.160	3.012	40	2.021	2.704
14	2.145	2.977	50	2.009	2.678
15	2.131	2.947	60	2.000	2.660
16	2.120	2.921	100	1.984	2.626
17	2.110	2.898	120	1.980	2.617
18	2.101	2.878			
19	2.093	2.861			

TABLE 8

The *F*-Distribution

If X has an F-distribution with m and n degrees of freedom, then table gives the value of x such that $\Pr(F < x) = .95$.

n	$m = 2$	$m = 3$	$m = 4$	$m = 5$	$m = 10$	$m = 15$	$m = 20$	$m = 30$	$m = 60$	$m = 120$
2	19.00	19.16	19.25	19.30	19.40	19.43	19.45	19.46	19.48	19.49
3	9.55	9.28	9.12	9.01	8.79	8.70	8.66	8.62	8.57	8.55
4	6.94	6.59	6.39	6.26	5.96	5.86	5.80	5.75	5.69	5.66
5	5.79	5.41	5.19	5.05	4.74	4.62	4.56	4.50	4.43	4.40
6	5.14	4.76	4.53	4.39	4.06	3.94	3.87	3.81	3.74	3.70
7	4.74	4.35	4.12	3.97	3.64	3.51	3.44	3.38	3.30	3.27
8	4.46	4.07	3.84	3.69	3.35	3.22	3.15	3.08	3.01	2.97
9	4.26	3.86	3.63	3.48	3.14	3.01	2.94	2.86	2.79	2.75
10	4.10	3.71	3.48	3.33	2.98	2.85	2.77	2.70	2.62	2.58
15	3.68	3.29	3.06	2.90	2.54	2.40	2.33	2.25	2.16	2.11
20	3.49	3.10	2.87	2.71	2.35	2.20	2.12	2.04	1.95	1.90
30	3.32	2.92	2.69	2.53	2.16	2.01	1.93	1.84	1.74	1.68
60	3.15	2.76	2.53	2.37	1.99	1.84	1.75	1.65	1.53	1.47
120	3.07	2.68	2.45	2.29	1.91	1.75	1.66	1.55	1.43	1.35

TABLE 9

Brief Table of Integrals

a, *b*, *c*, *m*, *n* represent constants;
C represents the arbitrary constant of integration.

Perfect Integral

$$\int dx = x + C$$

Multiplication by Constant

$$\int n\, dx = nx + C$$

$$\int nf(x)\, dx = n\int f(x)\, dx$$

$$\int f(nx)\, dx = \frac{1}{n}\int f(u)\, du$$

where $u = nx$

Addition

$$\int [f(x) + g(x)]dx = \int f(x)\, dx + \int g(x)\, dx$$

Powers

$$\int x\, dx = \frac{x^2}{2} + C$$

$$\int x^n\, dx = \frac{x^{n+1}}{n+1} + C \text{ if } n \neq -1$$

$$\int x^{-1}\, dx = \ln|x| + C$$

Polynomials

$$\int (a_n x^n + a_{n-1} x^{n-1} + \cdots + a_2 x^2 + a_1 x + a_0) dx =$$

$$\frac{a_n x^{n+1}}{n+1} + \frac{a_{n-1} x^n}{n} + \cdots + \frac{a_2 x^3}{3} + \frac{a_1 x^2}{2} + a_0 x + C$$

Substitution

$$\int f(u(x))\, dx = \int f(u) \frac{dx}{du} du$$

For example:

$$\int x f(x^2 + a)\, dx = \int x f(u) \left(\frac{1}{2x} \right) du = \frac{1}{2} \int f(u)\, du$$

where $u = x^2 + a$.

Integration by Parts

$$\int u\, dv = uv - \int v\, du$$

Note: The arbitrary constant of integration C will not be explicitly listed in the integrals that follow, but it must always be remembered.

Trigonometry

$$\int \sin x \, dx = -\cos x$$

$$\int \cos x \, dx = \sin x$$

$$\int \tan x \, dx = \ln|\sec x|$$

$$\int \sec x \, dx = \ln|\sec x + \tan x|$$

$$\int \sin^2 x \, dx = \frac{x}{2} - \frac{\sin 2x}{4}$$

$$\int x \sin x \, dx = \sin x - x \cos x$$

$$\int x^2 \sin x \, dx = -x^2 \cos x + 2x \sin x + 2 \cos x$$

$$\int \cos^2 x \, dx = \frac{x}{2} + \frac{\sin 2x}{4}$$

$$\int x \cos x \, dx = \cos x + x \sin x$$

$$\int \sin x \cos x \, dx = \frac{\sin^2 x}{2}$$

$$\int \sin^m x \, dx = -\frac{\sin^{m-1} x \cos x}{m} + \frac{m-1}{m} \int \sin^{m-2} x \, dx$$

$$\int \arcsin x \, dx = x \arcsin x + \sqrt{1 - x^2}$$

$$\int \arctan x \, dx = x \arctan x - \frac{\ln(1 + x^2)}{2}$$

Exponential Functions and Logarithms

$$\int e^x \, dx = e^x$$

$$\int x e^x \, dx = x e^x - e^x$$

$$\int x^2 e^x \, dx = x^2 e^x - 2x e^x + 2e^x$$

$$\int a^x \, dx = \frac{a^x}{\ln a}$$

$$\int e^x \cos x \, dx = \frac{e^x (\sin x + \cos x)}{2}$$

$$\int \ln x \, dx = x \ln x - x$$

$$\int x \ln x \, dx = \frac{x^2 \ln x}{2} - \frac{x^2}{4}$$

$$\int x^2 \ln x \, dx = \frac{x^3 \ln x}{3} - \frac{x^3}{9}$$

$$\int_{-\infty}^{\infty} e^{-x^2/2} \, dx = \frac{1}{\sqrt{2\pi}}$$

Integrals involving $ax^2 + bx + c$

For this section, let $D = b^2 - 4ac$. These integrals can be simplified by substituting $u = x + b/2a$:

$$ax^2 + bx + c = \frac{4a^2 u^2 - D}{4a}$$

(1) $\qquad$ Let $y = \int \dfrac{1}{ax^2 + bx + c} dx$

If $D < 0$: $y = \dfrac{2}{\sqrt{-D}} \arctan\left(\dfrac{2ax + b}{\sqrt{-D}}\right)$

$$\text{If } D > 0: \; y = \frac{1}{\sqrt{D}} \ln \left| \frac{2ax + b - \sqrt{D}}{2ax + b + \sqrt{D}} \right|$$

$$\text{If } D = 0: \; y - \frac{2}{2ax + b}$$

Specific examples of form (1) include:

$$\int \frac{1}{1 + x^2} \, dx \;\; = \;\; \arctan x$$

$$\int \frac{1}{1 - x^2} \, dx \;\; - \;\; \frac{1}{2} \ln \left| \frac{1 + x}{1 - x} \right|$$

$$\int \frac{1}{m^2 + n^2 x^2} \, dx \;\; = \;\; \frac{1}{mn} \arctan \left(\frac{nx}{m} \right)$$

$$\int \frac{1}{m^2 - n^2 x^2} \, dx \;\; = \;\; \frac{1}{2mn} \ln \left| \frac{m + nx}{m - nx} \right|$$

(2) $\qquad$ Let $y = \int \dfrac{1}{\sqrt{ax^2 + bx + c}} dx$

$$\text{If } a > 0: y = \frac{1}{\sqrt{a}} \ln \left| 2 \sqrt{a(ax^2 + bx + c)} + 2ax + b \right|$$

$$\text{If } a < 0 \text{ and } D > 0: y = \frac{-1}{\sqrt{-a}} \arcsin \left(\frac{2ax + b}{\sqrt{D}} \right)$$

(provided $|2ax + b| < \sqrt{D}$).

Specific examples of form (2) include:

$$\int \frac{1}{\sqrt{1-x^2}}\, dx = \arcsin x$$

$$\int \frac{1}{\sqrt{1+x^2}}\, dx = \ln(x + \sqrt{1+x^2})$$

$$\int \frac{1}{\sqrt{x^2-1}}\, dx = \ln(x + \sqrt{x^2-1})$$

$$\int \frac{1}{\sqrt{m^2-n^2x^2}}\, dx = \frac{1}{n}\arcsin\left(\frac{nx}{m}\right)$$

$$\int \frac{1}{\sqrt{n^2x^2+m^2}}\, dx = \frac{1}{n}\ln\left|\frac{nx}{m} + \sqrt{1+\frac{n^2x^2}{m^2}}\right|$$

(3)   Let $y = \int \sqrt{ax^2+bx+c}\, dx$

$$y = \frac{2ax+b}{4a}\sqrt{ax^2+bx+c} +$$

$$\left(\frac{4ac-b^2}{8a}\right)$$

$$\int \frac{1}{\sqrt{ax^2+bx+c}}\, dx$$

Specific examples of form (3) include:

$$\int \sqrt{1-x^2}\, dx = \frac{\arcsin x + x\sqrt{1-x^2}}{2}$$

$$\int \sqrt{1+x^2}\, dx = \frac{x\sqrt{1+x^2} + \ln|x + \sqrt{1+x^2}|}{2}$$

$$\int \sqrt{x^2 - 1}\, dx = \frac{x\sqrt{x^2 - 1} - \ln|x + \sqrt{x^2 - 1}|}{2}$$

$$\int \sqrt{m^2 - n^2 x^2}\, dx =$$
$$\frac{m^2}{2n} \left[\arcsin\left(\frac{nx}{m}\right) + \frac{nx}{m}\sqrt{1 - \left(\frac{nx}{m}\right)^2} \right]$$

$$\int \sqrt{m^2 + n^2 x^2}\, dx =$$
$$\frac{m^2}{2n} \left[\left(\frac{nx}{m}\right)\sqrt{1 + \left(\frac{nx}{m}\right)^2} + \right.$$
$$\left. \ln\left|\frac{nx}{m} + \sqrt{1 + \left(\frac{nx}{m}\right)^2}\right| \right]$$